D1759844

Quantum Physics for Beginners

Breaking Down Complex Quantum Theories into Easy-to-Understand Concepts and Practical Applications

By Alex Logman

Table Of Content

Introduction

The Importance of Quantum Physics

In today's interconnected world, an understanding of quantum physics might not seem immediately pressing. Yet, despite its abstract and elusive nature, this branch of science is deeply intertwined with our reality. Here's why.

Deconstructing the Quantum Myth

For many, quantum physics is confined to high-end laboratories or intricate thought experiments. But let's shatter that illusion. The realm of quantum isn't just about esoteric experiments; it's fundamental to our understanding of the universe. Why, you ask? Let's unpack this.

The Underlying Fabric of Reality

Every bit of matter around us is composed of atoms. These atoms aren't just static balls of energy; they're buzzing hives of activity, with electrons leaping between energy levels, particles flashing in and out of existence, and forces that boggle the mind. It's a chaotic dance on scales so tiny that they're almost unimaginable. And what dictates the rules of this dance? Quantum mechanics.

Quantum: The Unsung Hero of Modern Tech

Fast forward from these atomic playgrounds to our daily life. The smartphone in your pocket, the satellite navigating your car, even the lasers at the supermarket checkout—all these owe their functioning to our understanding of quantum principles. It's not just about creating smaller, faster transistors (though that's crucial).

It's about understanding materials at their core and crafting innovations previously thought impossible.

How to Navigate This Book

Diving into the vast ocean of quantum physics can be as exhilarating as it is daunting. But fear not. This book is your compass, your guide, and most importantly, your companion in this exploratory journey. Here's how to make the most of it:

1. Setting Your Pace

While quantum physics is complex, your learning journey doesn't have to be. Some chapters are denser than others, and it's perfectly okay to spend more time on areas that captivate or challenge you. Remember, it's not about speed, but comprehension. Slow down when needed, revisit earlier sections, or even jump ahead if curiosity gets the better of you.

2. Interconnecting Ideas

Quantum physics isn't a series of isolated concepts but a web of interrelated ideas. As you progress, you'll notice themes that echo and reinforce one another.

Embrace these connections. They'll deepen your understanding and offer a holistic view of the quantum realm.

3. Applying the Theoretical

Every chapter comes with practical applications, highlighting how these abstract concepts manifest in the real world. Engage with them, relate them to the tech enthusiast's sphere or any domain you belong to. This contextualizes quantum theory, making it tangible and relevant.

4. Exploring Beyond

In this book, I've provided foundational knowledge. But the quantum universe is vast. The "Further Reading and Resources" section in the appendix will guide you towards deeper explorations. Don't hesitate to dive into those additional resources; they offer enriching extensions to our discussions.

5. Engage in Discussions

Quantum physics thrives in discourse. Engage with fellow readers, join online forums, attend related seminars, or simply discuss with friends. Different perspectives and questions will illuminate facets of quantum mechanics you might not have considered.

Chapter 1

The Classical Physics Landscape

Newtonian Mechanics

A Glimpse Back in Time

The universe has always beckoned to us with its mysteries. Long before quantum physics was even a glint in the eye of science, classical mechanics paved the way for our understanding of physical phenomena. At the heart of classical mechanics lies the work of one Sir Isaac Newton.

The Birth of Newtonian Physics

In the late 17th century, Newton's groundbreaking 'Philosophiæ Naturalis Principia Mathematica' was published. In it, he introduced the three fundamental laws of motion that became the bedrock of classical mechanics.

1. **Newton's First Law (Inertia):** An object remains in its state of rest or uniform motion in a straight line unless acted upon by an external force. This concept, though simple, toppled millennia of misconceptions. It wasn't 'natural' for objects to come to a stop, as was previously thought, but rather, an external force (like friction) caused it.

2. **Newton's Second Law (F = ma):** The force acting on an object is equal to the mass of that object times its acceleration. This equation provided a quantitative method to predict how objects move under various forces.
3. **Newton's Third Law (Action and Reaction):** For every action, there's an equal and opposite reaction. If you push against a wall, it pushes back with equal force.

Implications and Triumphs

These laws, collectively, shaped the world of physics. From the rotation of planets to the trajectory of projectiles, Newtonian mechanics could predict it all with astonishing accuracy. Engineers, architects, and countless other professionals relied on these principles to build bridges, design vehicles, and chart out space missions.

Electromagnetism

When we think of the fundamental forces that shape our universe, our thoughts often soar to the vast realms of space, picturing celestial bodies and cosmic phenomena. However, there's a force right here on Earth, deeply embedded in our daily lives, yet often overlooked - electromagnetism. From the enchanting hues of the Northern Lights to the touchscreens we swipe every day, the touch of electromagnetism is ubiquitous.

Glimpses from the Past: Nature's Eerie Tales

Long before the intricacies of electromagnetism were unraveled, humankind noticed peculiar phenomena. Ancient Greeks witnessed a strange attraction when rubbing amber against animal fur. This 'static electricity' was akin to alchemy in their eyes. Think about the last time you experienced a shock touching a doorknob on a dry winter day. That's static electricity in action, a phenomenon that intrigued our ancestors and led them on a journey of discovery.

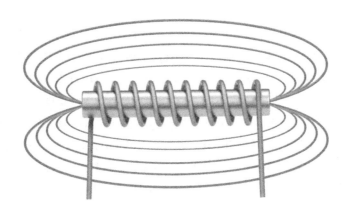

Sir William Gilbert: Charting the Unknown

Gilbert's pioneering work presented Earth as a colossal magnet. In modern terms, consider the compass apps in smartphones. The same principle that allowed ancient sailors to navigate vast oceans using magnetic compasses now enables us to find our way through labyrinthine city streets.

Gilbert's research set the stage for this, emphasizing the magnetic aura of our planet and its applications.

Hans Christian Oersted: Serendipity in Science

Picture Oersted's amazement when a routine demonstration unveiled the intimate link between electricity and magnetism. This discovery finds practical resonance in the workings of electric fans, where electricity running through coils produces a magnetic field, rotating the blades. That gentle breeze on a hot summer day has its roots in Oersted's revelation.

Faraday's Masterstroke: Unveiling Secrets of the Universe

Faraday's immersion in electromagnetic induction has contemporary echoes. Today, wireless charging pads for smartphones and electric toothbrushes are applications of Faraday's discovery. When you place your phone on a charging pad, and it powers up without any physical connection, it's Faraday's principles in action. These devices harness changing magnetic fields to generate electric currents, eliminating the need for cords.

James Clerk Maxwell: The Visionary Theorist

Maxwell's illumination of electromagnetism's symphony is groundbreaking. His predictions about electromagnetic waves find practical applications in everyday gadgets. Your microwave oven, for instance, uses these waves to heat food. When you tune into your favorite radio station or use Wi-Fi, you are accessing different parts of the electromagnetic spectrum Maxwell envisioned.

The Electromagnetic Spectrum: The Universe's Multifaceted Canvas

This spectrum is a vast continuum that touches our lives daily. Think about the last time you went for a sunbath. While you soaked in the sun's warmth, you were also absorbing infrared rays. If you stayed out too long and got sunburnt, you felt the effects of ultraviolet rays. Every time you use a remote control to switch channels or detect obstacles in the dark using night vision goggles, you're using infrared technology. On the other end, X-rays, which let doctors peer into our bodies, are also part of this spectrum. The colors we see, the warmth we feel, and the unseen signals that connect our devices - they all exist within this spectrum.

The Geomagnetic Shield: Earth's Silent Sentry

Our planet's magnetic field is more than just an academic interest. Consider the beautiful Aurora Borealis or Northern Lights. These mesmerizing displays result from solar winds interacting with Earth's magnetosphere. But beyond the beauty, the magnetosphere also acts as Earth's guardian, shielding us from harmful solar radiation. Astronauts on space missions, away from this protective barrier, need specially designed spacesuits and habitats to protect them from space radiation.

The Modern Era: Electromagnetism at the Heart of Progress

Today's technological marvels are a testament to the power of electromagnetism. Consider Magnetic

Resonance Imaging (MRI). It's a non-invasive technique that gives unparalleled views of soft tissues in our body, aiding in the early diagnosis of countless conditions. Or think about data transfer through fiber-optic cables, which employ light (an electromagnetic wave) to transmit information at lightning speeds. Each time we send a photo or stream a video, we're capitalizing on the principles of electromagnetism.

Electromagnetism, once a mysterious force, now pulses through the veins of our civilization. It lights up our homes, connects us across vast distances, powers our vehicles, and even safeguards our health. As we flip switches, tap screens, or simply bask in sunlight, we experience the profound legacies of those who unraveled the secrets of electromagnetism. Their curiosity has shaped our world, turning enigmas into conveniences and bringing the cosmos closer to our fingertips.

Thermodynamics

The term "thermodynamics" might seem abstract or esoteric to many, but its principles govern so much of what we experience daily. At its heart, thermodynamics is the study of energy, its transformations, and its interplay with matter. Imagine, if you will, a ballet, with energy and matter as the lead dancers, and the laws of thermodynamics as the choreography, directing their movements in the grand performance that is our universe.

In our homes, we've all boiled water for our morning coffee or tea. That process, as simple as it seems, is an embodiment of thermodynamics.

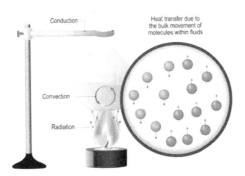

The heat from the stove is energy transferred to the water, increasing its temperature until it transforms into steam. This everyday ritual exemplifies two fundamental concepts of thermodynamics: energy transfer and phase transitions.

However, to fully appreciate thermodynamics, we need to delve a bit deeper and recognize its core principles. Historically, the development of thermodynamics was driven by the need to improve the efficiency of steam engines during the industrial revolution.

Pioneers like Sadi Carnot began asking foundational questions about the limits of energy conversion, leading to insights that have since transcended the realm of steam engines.

One such insight is the concept of energy conservation, encapsulated in the First Law of Thermodynamics. It posits that energy cannot be created or destroyed, only transformed. Whether it's the food we consume being metabolized into energy to power our bodily functions, or the electricity that charges our devices, energy's journey is one of constant transformation, but its total amount remains conserved.

Yet, it's the Second Law of Thermodynamics that often captures the most attention, primarily because of its philosophical implications about the nature of our universe. It states that the total entropy, or disorder, of an isolated system will always increase over time. In layman's terms, things tend to move from order to chaos. This might seem pessimistic or even nihilistic.

However, it's this very principle that drives processes essential to life. For example, our cells derive energy from the breakdown of glucose, a process fueled by the increase in entropy.

But thermodynamics isn't just about overarching laws. It's also about understanding specific properties of substances, like temperature, volume, and pressure, and their relationships. Consider the air conditioner that cools your home during summer.

Its operation is rooted in the phase transitions of refrigerants, substances that absorb and release heat under varying pressures and temperatures. By manipulating these conditions, the air conditioner cools the air, offering respite from the sweltering heat.

As our understanding of thermodynamics evolved, so did its applications, extending far beyond steam engines and home appliances.

Modern thermodynamics intersects with quantum mechanics, leading to the emerging field of quantum thermodynamics, which holds promise for next-generation technologies.

Concluding, thermodynamics, with its intricate interplay of energy and matter, dictates much of the world around and within us. From the beverages we warm up or cool down to the cells powering our bodies, to the vast stellar phenomena in space, the principles of thermodynamics are ever at work.

Chapter 2

The Quantum Revolution

The Ultraviolet Catastrophe

The evolution of science is rife with twists, turns, and surprises. Every once in a while, a discovery or observation emerges that challenges the existing paradigm and forces us to rethink everything we thought we knew. In the world of physics, the 'Ultraviolet Catastrophe' was one such seminal moment that signaled the end of classical thinking and ushered in the era of quantum mechanics.

Setting the Stage: The World of Classical Physics

At the turn of the 20th century, the world of physics was deeply entrenched in classical theories. These theories, which described the motions of planets and the behavior of heat and light, seemed nearly complete.

The equations set forth by Newtonian mechanics and Maxwell's electromagnetism had been successful in explaining a vast array of phenomena. It was an era where the universe seemed orderly, deterministic, and well-understood.

In this world, there existed the notion of a 'blackbody' - an idealized physical body that absorbs all incident electromagnetic radiation. When heated, blackbodies emit radiation in a spectrum that is determined only by their temperature. By the late 1800s, physicists were trying to describe this radiation spectrum in detail using classical physics.

The Rayleigh-Jeans Catastrophe

Lord Rayleigh and Sir James Jeans came up with a law based on classical electromagnetism and thermodynamics, aiming to describe the intensity of radiation emitted by a blackbody as a function of frequency. Their equation worked quite well for lower frequencies, fitting the experimental data.

However, as they ventured into higher frequencies, specifically the ultraviolet region, the predictions of their equation suggested that the blackbody would emit radiation with infinitely increasing energy. Such a scenario was termed as the 'Ultraviolet Catastrophe', signifying the catastrophic failure of classical physics in predicting a phenomenon that simply didn't make physical sense.

Planck's Leap into the Quantum

Enter Max Planck, a stalwart of classical physics, who was deeply troubled by this divergence between theory and observation. In an act of desperation, he decided to make a radical assumption. Instead of considering energy as something that can take on any value in a continuous manner, he hypothesized that energy can only be exchanged in discrete units or 'quanta'. In essence, energy became 'pixelated'.

This notion was groundbreaking. Energy, instead of being like a sliding scale, was more akin to individual steps on a staircase. Each step represented the smallest unit of energy, a quantum. Planck introduced a constant, now universally known as Planck's constant, to describe the size of these energy steps.

Using this new framework, Planck derived an equation that described the blackbody radiation spectrum. Not only did this new equation resolve the Ultraviolet Catastrophe, but it also fit perfectly with experimental data across all frequencies.

While Planck's introduction of quantized energy was initially met with skepticism, its implications were far-reaching. It set the stage for a series of discoveries and theories that would come to be collectively known as quantum mechanics.

A young Albert Einstein, inspired by Planck's work, soon introduced the idea of photons – quantized packets of light, further embedding the idea of quantization into the fabric of physics.

As the 20th century progressed, more phenomena that defied classical explanations were observed. Electrons in atoms, for instance, didn't orbit the nucleus as planets orbit the sun (as classical physics would predict). Instead, they existed in quantized energy states.

The Dual Nature of Light

When you imagine light, what do you see? Most of us envision a ray or a beam, something we have seen in countless illustrations and diagrams. This concept is deeply rooted in our understanding from centuries of study. Classical physics had predominantly treated light as waves—ripples that spread through the medium of the universe.

However, as the 20th century dawned, our understanding of light was about to undergo a dramatic transformation. The unfolding narrative of light's nature would reveal that it was not merely a wave, but also possessed particle-like properties.

This wave-particle duality of light challenged the very foundations of physics and reshaped our perception of reality.

The Age-Old Debate: Wave or Particle?

Historically, the nature of light was a matter of debate among scientists. Isaac Newton, in the 17th century, proposed that light consisted of particles, which he termed 'corpuscles'.

His argument was rooted in the observation that light travels in straight lines and can form sharp shadows, behaviors characteristic of particles.

Conversely, a contemporary of Newton, Christiaan Huygens, proposed a wave theory of light, emphasizing its ability to diffract and interfere—phenomena we typically associate with waves.

For a time, the wave theory seemed to gain prominence, especially with the work of Thomas Young in the early 19th century. Young's double-slit experiment showcased interference patterns, a direct indication of wave-like behavior. The wave model of light appeared triumphant.

Einstein's Quantum Leap with the Photoelectric Effect

Despite the triumphs of the wave theory, certain experiments defied explanation. One of the most striking was the photoelectric effect. When light of a certain frequency was shone onto a metal surface, it emitted electrons. Classical wave theory could not sufficiently explain why there was a threshold frequency below which no electrons were emitted, irrespective of the light's intensity.

Albert Einstein delved into this puzzle in 1905. He proposed that light, while behaving as a wave in some situations, can also act as a stream of particles or 'photons' in others.

These photons carry discrete packets or 'quanta' of energy, reminiscent of Planck's quantum idea. This energy is used to liberate electrons from the metal. Crucially, if the energy of individual photons is below a certain threshold, no electrons are emitted, no matter how many photons hit the surface.

Einstein's explanation was groundbreaking. It not only explained the photoelectric effect but also suggested that light had a dual nature. In some scenarios, it behaved as a wave, while in others, it manifested as particles.

Deepening the Mystery: Complementarity

While Einstein's photon hypothesis explained the photoelectric effect, it posed more questions. If light could be both a wave and a particle, when does it decide to be which?

Niels Bohr, a pioneering figure in quantum mechanics, introduced the principle of complementarity. He argued that light doesn't switch between being a wave or a particle. Instead, it possesses both wave-like and particle-like properties simultaneously. Which property manifests depends on the experiment or observation.

To understand this, consider the famous double-slit experiment. When photons are sent one by one through two slits, an interference pattern emerges on the screen behind, suggesting wave behavior.

Applications and Implications

This dual nature isn't just a philosophical conundrum; it has tangible implications in technology. Solar cells, for instance, operate based on the photoelectric effect. Light photons strike the cell, liberating electrons and producing electricity.

Moreover, our modern understanding of lasers, LED lights, and even the functioning of our eyes hinges on acknowledging both wave-like and particle-like characteristics of light. In medical fields, treatments like photodynamic therapy for cancer utilize the particle nature of light to activate certain drugs selectively.

The wave-particle duality of light is more than a mere scientific oddity; it's a profound insight into the nature of the universe. It teaches us that our classical notions, however intuitive, might only represent a slice of a more complex reality.

The Bohr Model of the Atom

In our ever-expanding quest to understand the universe, no topic has perplexed and fascinated us quite like the behavior of the tiniest of particles. Among these minute wonders, the electron stands out as a particularly intriguing player in the quantum realm. But why? What makes the electron so indispensable to our understanding of quantum mechanics and, by extension, the very fabric of reality?

The Enigmatic Electron: A Historical Context

When J.J. Thomson first discovered the electron in 1897, the scientific community was thrilled. Here was a subatomic particle, even smaller than an atom, and it held secrets that would soon revolutionize physics. However, as with many aspects of quantum mechanics, the deeper we delved into the nature of the electron, the more mysterious it became.

Early on, we learned that electrons orbit the nucleus of an atom.

This raised questions: Why don't electrons simply crash into the nucleus, attracted by its positive charge? The answer would come from quantum mechanics, and it would be unlike anything we had heard before.

Orbitals, Not Orbits

While it's easy to picture electrons circling the nucleus in neat, defined orbits (much like planets around the sun), this is not the case. Instead, electrons exist in "orbitals," regions around the nucleus where they are most likely to be found. These orbitals aren't straightforward paths but rather fuzzy clouds of probability.

To elucidate, consider a buzzing cloud of bees around a beehive. You cannot predict where any single bee might be at a given time, but you can say with certainty that most bees will be near the hive. Similarly, an electron's position isn't determined until we observe it.

The Principle of Quantum Leaps

Electrons in an atom don't exist in just any energy state. They occupy specific energy levels. When they transition between these levels, they do so instantaneously, without passing through
the space in between. This phenomenon is known as a "quantum leap."

Imagine standing at the base of a staircase. In our everyday experience, you can climb the stairs step by step. But in the quantum realm, it's as if you're magically transported from the bottom step to a higher one without touching the ones in between.

Electron Spin: A Quantum Twist

Electron behavior becomes even more enigmatic when we consider "spin." Spin isn't about an electron physically rotating; it's an intrinsic form of angular momentum.

But, intriguingly, electrons can only have one of two spin values: up or down. This binary nature of electron spin underpins many quantum phenomena, including the Pauli exclusion principle, which states that no two electrons in an atom can have the same set of quantum numbers.

Understanding electron spin has led to revolutionary applications. Modern technologies, like Magnetic Resonance Imaging (MRI) in medical fields, leverage the principles of electron spin to produce detailed images of the human body.

Just as light exhibited both wave-like and particle-like properties, so too did electrons. The famous double-slit experiment, when conducted with electrons, showed interference patterns reminiscent of waves, even when electrons were fired one at a time. This reinforced the idea that the quantum world defies classical intuition.

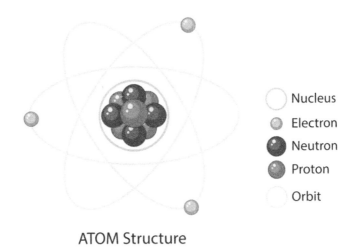

Nucleus
Electron
Neutron
Proton
Orbit

ATOM Structure

Chapter 3

Key Principles of Quantum Physics

Wave-Particle Duality

In our quest to decode the enigmatic realm of quantum physics, we've encountered ideas that challenge conventional wisdom. But if there's one principle that stands as a cornerstone of quantum mechanics, illuminating yet confounding our understanding, it's the concept of wave-particle duality.

For centuries, the dominant worldview segregated waves and particles into distinct categories. Waves, like those in a pond or an oscillating string, were described as continuous disturbances spreading across a medium. Particles, on the other hand, were considered tiny, localized points of matter with definite paths. Then, as we peeled back the layers of the atomic and subatomic worlds, this neat categorization began to blur, leading to a revolution in our understanding.

The Essence of Duality

Imagine a tranquil lake. If you were to throw a stone into it, concentric ripples would emerge, spreading outward. Now, think of a tiny grain of sand, visible, tangible, and precise in its location.

For the longest time, scientists believed that waves and particles, much like the ripples and the grain, were mutually exclusive. Yet, quantum mechanics proposed something audacious: particles can behave like waves, and waves can exhibit particle-like characteristics.

Double-Slit Experiment: A Quantum Enigma

To illustrate wave-particle duality, we can turn to the iconic double-slit experiment, an experiment that remains one of the most profound in all of physics. Here's a simplified version of how it goes:

Shoot particles (like electrons) at a barrier with two slits, and place a screen behind the barrier. Now, classically, we'd expect

each electron, acting as a particle, to pass through one slit and hit the screen, producing two bands corresponding to the slits. Instead, when observed, an interference pattern emerged on the screen, much like waves combining or canceling each other out. It was as if each electron, instead of going through one slit, went through both as a wave and interfered with itself.

But the story becomes even stranger. When scientists tried to "catch" the electron in the act, determining which slit it went through, the interference vanished! It was as if the very act of observation collapsed the wave nature, forcing the electron to "choose" a particular path. This wasn't just a problem of measurement; it was a fundamental aspect of quantum reality.

The Uncertainty Principle - Grasping the Elusive Quantum Realm

Within the tapestry of quantum mechanics, certain threads stand out as peculiar, making us question our understanding of reality. The Heisenberg Uncertainty Principle is one such thread.

Setting the Stage: An Unsettling Observation

Werner Heisenberg, a pivotal figure in the early development of quantum mechanics, introduced the Uncertainty Principle in 1927. As scientists probed the quantum realm, they realized that certain pairs of properties (like position and momentum) couldn't be precisely determined simultaneously. The more accurately you tried to measure one, the less accurately you could measure the other.

The Heart of the Matter: Unveiling the Principle

At a glance, the Uncertainty Principle can be described as the idea that certain pairs of physical properties, like position (x) and momentum (p), are intrinsically linked in such a way that they can't both be precisely measured at the same time. Mathematically, it's expressed as:

$$\Delta x * \Delta p \geq h/4\pi$$

Here, Δx is the uncertainty in position, Δp is the uncertainty in momentum, and h is Planck's constant.

It's crucial to understand that this isn't a limitation of our instruments or measurement techniques. Instead, it's a fundamental aspect of quantum systems. In essence, the quantum world is inherently 'fuzzy'.

Ripples in the Pond: Implications of the Uncertainty

On a macroscopic scale, the effects of the Uncertainty Principle are negligible. But as we venture into the quantum realm, it becomes a dominant player.

Electrons in atoms, for instance, don't have well-defined orbits as classical physics would predict. Instead, they exist in 'clouds' or probability distributions around the nucleus. This nebulous existence can be directly attributed to the Uncertainty Principle.

The Principle also has profound implications for the very nature of reality. It suggests that the universe, at its most fundamental level, is indeterminate. Rather than a clockwork universe where future events can be precisely predicted if we know the current conditions, the quantum universe is one of probabilities and tendencies.

Applications and Insights: Harnessing the 'Fuzziness'

While the Uncertainty Principle might seem abstract, it has concrete implications in various fields. For instance, in electronics, the behavior of semiconductors and transistors at nano scales is heavily influenced by quantum uncertainties.

Understanding these principles allows us to innovate and design more efficient electronic devices.

Another fascinating application is in the field of Quantum Cryptography. The Uncertainty Principle can be employed to develop secure communication channels. If a third party tries to intercept the quantum data being exchanged, the very act of measuring will disturb the quantum system, alerting the parties involved about a potential eavesdropper.

Superposition and Entanglement

The world of quantum physics is awash with phenomena that are undeniably strange, consistently defying our everyday intuitions.

Among these, the phenomenon of entanglement stands out, not only due to its sheer peculiarity but also because of its profound implications on our understanding of nature and the very fabric of reality.

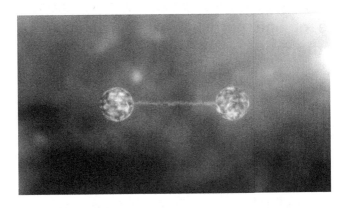

A Brief Historical Prelude

The story of entanglement begins with a debate, specifically, the heated exchanges between two giants of 20th-century physics:

Albert Einstein and Niels Bohr. While Bohr was an ardent proponent of quantum mechanics and its inherent probabilistic
nature, Einstein was skeptical. He believed that the universe must have a deterministic underpinning, famously declaring, "God does not play dice with the universe." To challenge the quantum theory, Einstein, along with colleagues Podolsky and Rosen, proposed a thought experiment in 1935, which has since become known as the EPR paradox. This experiment was an attempt to reveal the "incompleteness" of quantum mechanics.

The Core Idea: Spooky Action at a Distance
At the heart of the EPR paradox was the concept of entanglement. Imagine two particles that have interacted with each other and then are separated,
traveling vast distances apart. According to quantum mechanics, these particles remain entangled, meaning the state of one particle is inexorably linked to the state of the other, no matter the distance separating them.

If one were to measure the state of one particle, the state of the other particle would be instantly determined, even if it were light-years away. This instant correlation seemed to violate the principle that information cannot travel faster than light, leading Einstein to call it "spooky action at a distance."

Bell's Theorem and the Validation of Entanglement
In the 1960s, physicist John Bell formulated a set of

inequalities that could test the validity of the spooky action concept. If these inequalities were violated in experiments, it would confirm the intrinsic non-locality of quantum mechanics and give credence to the idea of entanglement.

Bell's challenge was taken up by experimenters, most notably by Alain Aspect in the early 1980s. Using pairs of entangled photons, Aspect and his team found results that violated Bell's inequalities, thus supporting the reality of quantum entanglement. These experiments didn't prove faster-than-light communication; instead, they indicated a profound interconnectedness in the quantum realm, an inherent non-separability.

Practical Implications: The Quantum Revolution

Beyond its philosophical implications, entanglement is paving the way for revolutionary technologies. Consider quantum computing, where the power of qubits (quantum bits) arises from their ability to be in superpositions and their entangled states.

This enables quantum computers to process a vast amount of information simultaneously, potentially solving problems deemed insurmountable for classical computers.

Another promising area is quantum cryptography. By using entangled photons, it's possible to create communication channels that are inherently secure. Any attempt to eavesdrop on the communication would disturb the entangled state, alerting the communicating parties to the intrusion.

Chapter 4
Quantum States and Wavefunctions

The Concept of Quantum States

At the heart of quantum mechanics lies an idea so profound, so fundamentally different from classical physics, that it reshapes our understanding of reality: the quantum state. It's a concept that has both intrigued and puzzled scientists, philosophers, and enthusiasts alike. Let's dive deep into this captivating world, taking a detailed and comprehensive look at the nature of quantum states and the mysteries they hold.

To start our journey into the realm of quantum mechanics, we must first recognize that our classical intuition, derived from everyday experiences, often falters in the quantum realm.

Understanding Quantum States

In classical physics, the state of an object is defined by its position and velocity at any given time. For instance, if you knew the exact location and speed of a baseball, you could predict its future trajectory.

But quantum mechanics doesn't operate on these straightforward principles. Instead of a definite position or velocity, particles are described by a quantum state, often represented by a wavefunction.

This state contains all the information about the particle, but instead of clear-cut values, we get probabilities.

Imagine a particle inside a box. In classical physics, we could say exactly where the particle is and how fast it's moving. In quantum physics, however, the particle doesn't have a specific location until measured. Instead, there's a probability cloud of where we might find the particle. This cloud is densest where the likelihood of finding the particle is highest and thinnest where it's least likely.

Wavefunction: The Mathematical Persona of Quantum States

The wavefunction, usually represented by the Greek letter ψ

(psi), is a mathematical function that describes the probability distribution of a particle. When we square the magnitude of the wavefunction at any point in space, it gives the probability of finding the particle at that location.

Consider our particle in a box again. The shape and size of the probability cloud (our wavefunction) can change based on various factors, such as external forces or interactions with other particles. And here's the truly baffling part: until we measure the particle's position, it exists in a superposition of all possible positions.

It's as if the particle is simultaneously everywhere and nowhere in the box.

Collapse of the Wavefunction

One of the most debated aspects of quantum mechanics is the act of measurement. It's said that when a quantum state is observed or measured, the wavefunction "collapses" to a single value.

Using our particle in a box example, once we measure its position, the particle will be found at a specific location, and the probability cloud collapses to that point.

But why does this happen? Why does the act of observation force a quantum system into a particular state? This question leads us to various interpretations of quantum mechanics, some of which suggest that the wavefunction doesn't really collapse, but rather splits into multiple branches, leading to the concept of multiple realities or "many worlds".

The Role of Wavefunctions

The intriguing landscape of quantum mechanics, with its plethora of novel ideas and paradigms, finds its foundation in wavefunctions. Wavefunctions are not merely mathematical constructs; they play a pivotal role in shaping the quantum narrative, offering insights into the probabilistic nature of quantum entities.

Their importance cannot be understated, for they bridge the divide between abstract quantum theories and observable phenomena.

In this section, we aim to unravel the deep significance and the broad spectrum of applications of wavefunctions in the realm of quantum mechanics.

The Essence of Wavefunctions

At its core, a wavefunction provides a complete description of a quantum system. It encapsulates all the possible information about a quantum entity, be it a particle like an electron or even a more complex system. But unlike classical descriptions which grant definitive values (like exact positions and velocities), wavefunctions deal in probabilities.

When we talk about probabilities in quantum mechanics, it's essential to understand that these aren't born out of our ignorance or lack of precise instruments. Instead, they're intrinsic to the nature of quantum entities.

The wavefunction, represented by the symbol ψ (psi), tells us about these inherent probabilities.

Probabilistic Predictions and Reality

Let's envisage a simple scenario: a particle enclosed in a box. Classically, we would determine its position by looking inside. However, in quantum mechanics, until we make an observation,

the particle is described by a wavefunction that represents the probability of finding the particle at any point inside the box. The particle doesn't possess a definite position until observed.

When we talk about the probability associated with a wavefunction, we refer to the square of its magnitude at a particular point. This squared magnitude, $|\psi|^2$, provides the probability density for the particle's location.

However, a natural question arises: Why do quantum entities abide by probabilistic rules? The answer, while not entirely resolved, leans on the fundamental differences between the classical and quantum worlds. Quantum entities do not fit within our everyday experiences and thus defy our classical intuitions.

Wavefunctions and Interactions

In the world around us, interactions form the crux of dynamics. Similarly, in quantum mechanics, when two particles come close, their wavefunctions can interfere, leading to constructive (amplification) or destructive (cancellation) outcomes. This phenomenon is strikingly evident in the famous double-slit experiment, where particles like electrons, when passed through two slits, interfere with themselves, creating an interference pattern typical of waves.

Furthermore, particles can get entangled, a deeply quantum phenomenon where the wavefunctions of two particles become linked. An action on one particle instantly influences the state of the other, irrespective of the distance between them. This interplay of wavefunctions underscores the non-local nature of quantum mechanics, where actions in one location can have consequences elsewhere, defying our classical sense of separability.

Probability Distributions in Quantum Mechanics

In the quantum world, the notion of certainty takes a backseat, and in its place, probabilities govern the dynamics of particles. While we've touched upon wavefunctions as entities that describe the likelihood of finding a particle in a particular state, it's crucial to delve deeper into the idea of probability distributions, which form the essence of quantum predictions.

The Fabric of Probabilities

Imagine the vast tapestry of the universe, with every thread representing a potential path or state a particle can be in. While this imagery might seem overwhelming, it's this vast multitude of possibilities that quantum mechanics seeks to address through probability distributions.

A probability distribution is essentially a map, showing the likelihood of a particle existing in a particular state or position at any given moment.

It's like a weather forecast, predicting the chances of rain but on a quantum scale.

Instead of clouds and precipitation, we're concerned with particles like electrons, photons, and even larger entities under certain conditions.

Constructing Distributions from Wavefunctions

You may recall that the wavefunction, symbolized as ψ, offers a mathematical description of a quantum state.

The absolute square of this wavefunction, denoted as $|\psi|^2$, provides us with the probability density of finding the quantum entity in a specific location or state. Now, this 'density' is a crucial term to grasp. Rather than giving an outright probability, it offers a value that, when integrated over a region, yields the likelihood of the particle's presence in that region.

To bring this to life, let's consider an electron around an atom. If we were to map the probability density $|\psi|^2$ at every point around the atom, we'd observe regions where the density peaks, indicating higher chances of finding the electron there.

These peaks are what we commonly term as 'orbitals' in atomic physics. The unique shapes of s, p, d, and f orbitals are a direct consequence of the probability distributions derived from wavefunctions.

Interplay of Measurements and Probabilities

In classical mechanics, we're used to measuring quantities with definitive outcomes. If you were to toss a coin, the outcome is either heads or tails. However, in quantum mechanics, until a measurement is made, particles exist in a superposition of states.

This superposition isn't merely a limitation of our instruments but a fundamental feature of quantum entities.

When a measurement is made, the wavefunction collapses, and the particle is found in a specific state. This process is inherently probabilistic, governed by the aforementioned probability distributions.

If a particular outcome has a higher probability density, it's more likely but not guaranteed to be observed upon measurement.

Chapter 5

Quantum Operators and Observables

The Role of Operators

In the tapestry of quantum mechanics, there are threads that tie together all our observations, all the seemingly mystical behaviors of particles, and all the mathematical formalisms into a coherent understanding. Among these threads, quantum operators play a pivotal role. They are essential mathematical tools, acting as bridges that connect the abstract realm of quantum theory with the concrete outcomes of measurements in the lab. In this section, we will delve deeply into the concept of operators, their importance, and the insights they offer into the quantum realm.

Operators: The Quantum Mechanics' Command

Imagine for a moment you're watching a grand play. The stage is set, the actors (particles) are in their positions, ready to enact their roles. But for any play to proceed, there needs to be a director, someone to command actions, to guide the narrative. In the realm of quantum mechanics, operators take on this directorial role. They command the quantum system, asking it questions, and in response, the system provides answers in the form of measurable quantities, or observables.

An operator acts on a quantum state (described by wavefunctions, as discussed in the previous chapter) and provides valuable information about certain attributes of that state.

Whether we want to know about a particle's position, momentum, energy, or spin, there are specific operators associated with each of these observables.

Operators in Action: The Uncertainty Principle

One of the most profound and counterintuitive aspects of quantum mechanics is Heisenberg's uncertainty principle, which we touched upon in Chapter 3. To truly appreciate the depth of this principle, we need to understand the role of operators

Heisenberg's principle isn't just a statement about our lack of perfect knowledge; it arises from the very mathematical structure of the position and momentum operators. When these operators act on a quantum state, their non-commutativity (meaning the order in which they're applied matters) gives rise to intrinsic uncertainties in simultaneous measurements of position and momentum.

Real-World Implications: Quantum Tunnelling

To appreciate the importance of operators, consider the phenomenon of quantum tunnelling. An electron encountering a barrier, classically speaking, shouldn't pass through unless it has enough energy. But in the quantum realm, there's always a non-zero probability that the electron will 'tunnel' through, even if its energy is insufficient.

How does this happen? It's the momentum operator in action. When the momentum operator acts on the wavefunction of the electron near the barrier, it reveals a spread in possible momentum values. Some of these values allow the electron to tunnel through, manifesting the non-deterministic nature of quantum mechanics.

In many ways, understanding operators is like deciphering the notes on a sheet of music. Each note, each operator, tells a part of the story.

But when played together, they produce the symphony that is quantum mechanics. Operators give voice to particles, allowing them to reveal their attributes, their

behaviors, and their interactions.

As we continue our exploration into quantum mechanics, the role of operators will consistently be at the forefront, guiding our understanding and shaping our insights into the deepest secrets of the universe.

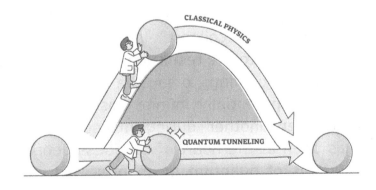

Measurement and Observables

Quantum mechanics, in all its fascinating complexity, remains grounded by one crucial fact: At some point, the predictions made by the theory have to correlate with the physical measurements we observe in our laboratories. After all, a theory, no matter how mathematically elegant, loses its relevance if it can't describe the natural world. In the heart of this bridge between the ethereal quantum world and our palpable reality lie the concepts of measurement and observables.

The Quantum Act of Measurement

Measurement in quantum mechanics is unlike anything in classical physics. Imagine a vast sea with countless waves, some tall, some short, all superimposed on each other. This sea is akin to a quantum system in a superposition of states.

When we measure this system, it's like freezing one of those waves in a snapshot, making it the reality at that moment. This act of measurement causes the quantum system to 'collapse' into a specific state, which corresponds to the result of our measurement.

Such behavior is baffling, to say the least. Why would a system with potentially infinite possibilities choose one state over another during measurement? This query has been at the heart of countless debates and interpretations of quantum mechanics, and while we don't have all the answers, what we do know is how to predict the probabilities of these outcomes.

The Dual Role of Wavefunctions in Measurement

Wavefunctions, as we've explored, provide a mathematical description of quantum states. But they play a dual role when it comes to measurements.

First, the absolute square of a wavefunction at a given point in space gives the probability density of finding a particle at that location.

Second, when an operator corresponding to an observable acts on this wavefunction, the resulting values give us insights into the probable outcomes of measurements.

Consider, for example, the simple act of measuring the position of an electron in a potential well. The wavefunction of the electron might be spread out across the well, suggesting it could be found anywhere inside. However, the act of measuring its position will yield a definitive result, say, "the electron is here." But if we repeat this experiment multiple times, we'll find the electron in different positions each time, in a manner consistent with the probability distribution given by the wavefunction's absolute square.

The Paradox of Schrödinger's Cat

To further understand the profound implications of quantum measurements, let's delve into the famous thought experiment of Schrödinger's cat. Imagine a cat inside a sealed box along with a radioactive atom, a detector, and a vial of poison. If the atom decays, the detector triggers the release of the poison, killing the cat.

But quantum mechanics suggests that until we

measure the state of the atom, it's in a superposition of decayed and undecayed states. Does this mean the cat is both alive and dead until we open the box and observe?

This paradox showcases the challenges and counterintuitive nature of quantum measurements.

Commutation and Uncertainty

Quantum mechanics, often regarded as nature's most enigmatic theory, is rich with phenomena that challenge our classical understanding of the world. Among these phenomena, the relationship between commutation and uncertainty stands out as one of the most mystifying.

The Concept of Commutation in Simple Terms

At its core, commutation in quantum mechanics revolves around the idea of sequencing. In our daily lives, the order in which we carry out certain tasks often doesn't affect the outcome.

For instance, whether you put on your socks before your shoes or vice versa, the outcome remains that you're ready for your day.

However, in the quantum world, the order can matter immensely. For certain pairs of quantum properties, measuring one before the other can yield vastly different results than if you swapped the order.

From Classical Confusion to Quantum Clarity: Position and Momentum

Let's delve into a tangible example: the intertwined relationship between a particle's position and its momentum (essentially its movement). In our classical world, we could, in theory, pinpoint a car's location and speed without any issues. But, quantum particles aren't so straightforward.

If we're extremely certain about a quantum particle's position, its momentum becomes more uncertain, and vice versa. This isn't a flaw in our instruments or techniques; it's a fundamental characteristic of the quantum world.

Implications of this Relationship: Quantum Tunneling

One of the most baffling outcomes of this relationship is the phenomenon of quantum tunneling. Imagine a hill, and you're pushing a boulder up this hill. If you don't push hard enough, the boulder won't make it over the hill. That's our classical understanding.

Now, translate this to the quantum world. An electron (our quantum equivalent of the boulder) approaches a barrier. Even if it seemingly doesn't have enough energy to cross this barrier (akin to the boulder not having enough push to get over the hill), there's still a chance it might just appear on the other side. This isn't because it got a sudden boost of energy; it's an inherent probability in the quantum realm.

Chapter 6
Quantum Dynamics

Time Evolution of Quantum States

Time stands as one of the most enigmatic concepts of our universe. The very fabric of reality is interwoven with its unceasing flow. When we transition from the macroscopic world, which adheres to classical physics, into the microscopic realm dominated by quantum mechanics, our understanding of time and its effects must adapt. One of the critical aspects of this adaptation is the time evolution of quantum states.

The Unfolding Symphony of Quantum Evolution

Every moment, particles in the quantum realm undergo subtle shifts and changes, reminiscent of the ebbs and flows in a musical symphony. These shifts, which represent the heart of quantum time evolution, are comparable to how instruments in an orchestra create a dynamic, ever-changing soundscape.

Imagine a violin section in an orchestra, where each player adjusts the pitch of their string simultaneously. This collective shift creates a change in the melody. Similarly, in a quantum system, particles like electrons may change their states in response to some external stimulus, resulting in a different quantum "melody."

The Quantum Landscape and Its Dynamics

Within the quantum realm, particles exist in a state of

probabilities. These probabilities dictate the likelihood of finding a particle in a particular state or position. Over time, these probability distributions shift and evolve, creating the dynamic tapestry of quantum mechanics.

Consider a school of fish swimming in the ocean. Initially, they might spread out, exploring a coral reef. However, the sudden appearance of a predator would immediately alter their distribution, with most clustering together for safety. This swift change in distribution is reminiscent of how quantum states can evolve when "disturbed" by an external factor.

Driving Forces Behind Quantum Time Evolution

The continuous dance of quantum states is orchestrated by several factors. These factors can be both external and internal,

each playing a pivotal role in the grand ballet of quantum dynamics.

1. **External Influences:** The world outside a quantum system can exert considerable influence over its time evolution.

Think of sunflowers in a field, which turn their heads to follow the sun's movement across the sky. This behavior, heliotropism, is a direct response to an external influence. In a similar vein, quantum states might evolve in reaction to external energies or fields.

2. **Internal Interactions:** Quantum systems, especially those comprising multiple particles, exhibit intricate internal interactions that drive

their time evolution.

Within a beehive, each bee communicates and interacts, leading to collective behaviors like swarming. Similarly, particles within a quantum system might interact, leading to collective quantum states or behaviors.

3. **Measurement and Observation:** The world of quantum mechanics introduces a peculiar phenomenon where the act of observation can dramatically alter a system's state.

Picture a classroom where students might act differently when the principal suddenly enters. The mere act of observation changes behavior. In quantum mechanics, merely observing or measuring a system can collapse its state, leading to specific outcomes.

The Schrödinger Equation: The Heartbeat of Quantum Dynamics

The Essence of the Schrödinger Equation

The Schrödinger equation, avoiding its intricate mathematical details, primarily describes how the quantum state of a physical system changes over time. Its beauty and complexity arise from its dual nature:

1. **Wave Functions and Probabilities:** Rather than deterministic predictions, the equation offers a probabilistic view. This marked a paradigm shift from the classical realm, where systems were governed by strict causality.

The quantum state, represented by the wave function, provides a probability distribution that indicates where a particle might be found upon measurement.

Example: Consider the act of shuffling a deck of cards. Before revealing any card, one can only estimate the position of each card, akin to the probabilistic nature of quantum states.

When a card is revealed, it's analogous to the act of measurement in quantum mechanics, collapsing the myriad possibilities into one observed reality.

2. **Interplay of Energy and Dynamics:** The Schrödinger equation beautifully interweaves the energy of a system with its dynamic evolution. It inherently suggests that the energy of a quantum system directly influences the manner in which its wave function evolves over time.

Example: Visualize a ballet dancer, whose movements (analogous to the wave function's evolution) are directly influenced by the music's rhythm (akin to the system's energy).

A change in tempo can lead to a completely different set of dance moves, just as varying energy scenarios can lead to distinct wave function evolutions.

Key studies, such as the one by Werner Heisenberg on quantum energy levels, further reinforced the intimate connection between energy and quantum dynamics.

3. **Embracing Indeterminacy:** The Schrödinger equation, much to the chagrin of many

physicists of the time, embraced non-deterministic outcomes. This was a significant departure from classical mechanics, where determinism was the norm.

Example: Think of tossing a coin. Even if we know every detail of how the coin was tossed – the force, the angle, the height – predicting its outcome remains inherently random. Similarly, even if we know a quantum system's initial state, predicting its exact future state becomes a game of probabilities.

Quantum Transitions and Decoherence

The realm of quantum mechanics, though rooted in complex mathematical formulations and abstract concepts, paints a fascinating picture of how subatomic particles dance, transition, and occasionally lose their rhythm in this grand cosmic ballet.

Quantum Transitions: The Ephemeral Leaps

Quantum transitions refer to the phenomenon where a quantum system transitions from one state to another. Unlike classical systems, where changes are typically continuous and gradual, quantum systems are characterized by discrete, quantized jumps between states.

1. **The Spectral Lines Enigma**: One of the most striking evidences of quantum transitions is found in the observation of atomic spectra. When atoms are excited, they emit or absorb light at specific frequencies, resulting in spectral

lines. These lines are nothing but visual manifestations of quantum transitions.

Example: The phenomenon is analogous to a multi-storied elevator, where the elevator can only stop at distinct floors, not in-between. Atoms, when energized, 'jump' between these energy levels, emitting or absorbing specific frequencies of light corresponding to the energy difference between the levels.

2. **Selection Rules and Forbidden Transitions**: Not all transitions are allowed in quantum systems. The term "selection rules" refers to the criteria that dictate which transitions are possible and which are forbidden. These rules are derived from underlying symmetries and conservation laws.

Example: Visualize a game of chess. There are specific rules on how each piece can move. A bishop, for instance, moves diagonally and cannot suddenly jump to an adjacent square like a knight. Similarly, quantum particles have their set of rules, defined by the inherent properties and constraints of the quantum world.

Decoherence: When Quantum Systems Lose Their Beat

Decoherence is a phenomenon that has intrigued and puzzled quantum physicists for decades. It refers to the process by which a coherent quantum superposition breaks down, typically due to interactions with the environment, leading to the emergence of classical behaviors from quantum systems.

1. **The Mystery of the Cat**: Erwin Schrödinger's famous thought experiment involving a cat that is simultaneously alive and dead until observed provides a vivid illustration of quantum superposition. However, in reality, such superpositions are incredibly fragile. Interactions with the external environment, even minimal, can disrupt these superpositions, a process termed decoherence.

Example: Consider a finely tuned musical instrument. When played in a quiet room, it produces clear and resonant tones. However, in a noisy environment, the instrument's sound can be drowned out and distorted. Similarly, the delicate superposition of quantum states gets 'distorted' or collapses due to environmental interactions.

2. **Quantum Computing and Decoherence**: Decoherence isn't just a theoretical concern; it poses practical challenges, especially in the emerging field of quantum computing. Quantum bits or qubits rely on maintaining superposition for quantum computations. Decoherence can disrupt these computations, leading to errors.

Example: Imagine a synchronized swimming team performing a routine. If even one swimmer gets out of sync, it can disrupt the entire performance. Similarly, in quantum computers, the loss of coherence can hamper the entire computational process.

Chapter 7

Quantum Mechanics and Atomic Structure

The Hydrogen Atom

In the landscape of atomic study, the hydrogen atom holds an esteemed position.

Comprising just a single proton and electron, it appears deceptively simple, yet its behavior, when viewed through the lens of quantum mechanics, was instrumental in revolutionizing our understanding of the atomic world.

Understanding the Quantum Hydrogen Atom

Bohr's model was groundbreaking, but it was merely the precursor to a more profound understanding brought about by quantum mechanics.

Enter the wavefunction, a mathematical construct that encapsulates all the information we can know about the state of a quantum system. For the hydrogen atom, the wavefunction, often denoted by ψ, provides a probabilistic description of the electron's location.

In this quantum perspective, electrons don't traverse defined paths around the nucleus. Instead, they exist in a cloud of probability, where the density of the cloud at any point gives the likelihood of finding the electron there.

This cloud forms the "electron orbitals" we often visualize today, from the spherical s-orbitals to the dumbbell-shaped p-orbitals.

Significance of the Quantum Model

The implications of the quantum model for the hydrogen atom were far-reaching:

1. **Discreteness of Energy Levels**: Quantum mechanics naturally explained the discrete spectral lines of hydrogen. Each spectral line corresponds to a transition between two energy states, and the energy of the emitted or absorbed photon is precisely the difference between these energy levels.

2. **Stability of Atoms**: The quantum model explained why electrons don't spiral into the nucleus. In the quantum realm, electrons can't have just any arbitrary energy; they're restricted to specific energy levels. The lowest of these levels, the ground state, provides a stable, lowest-energy configuration for the atom.

3. **Foundations for More Complex Atoms**: The principles applied to the hydrogen atom formed the foundation for understanding more complex atoms. Although the math becomes more intricate, the core concepts, such as quantized energy levels and probabilistic electron distributions, remain consistent.

Multi-Electron Atoms

After the profound exploration of the hydrogen atom,

the natural progression in the narrative of atomic physics is the understanding of atoms with more than one electron. These multi-electron atoms introduce layers of complexity, intrigue, and elegance to quantum mechanics.

The tantalizing dance of electrons in the hydrogen atom morphs into an elaborate ballet when additional electrons come into play.

The Complexity of Multiple Interactions

In a hydrogen atom, the quantum dance is relatively straightforward - we have one electron experiencing the pull of one proton. However, introduce another electron into this scenario, and the dynamics change drastically. In multi-electron atoms:

1. **Electron-Electron Interactions**: Each electron is not only attracted to the nucleus but also repelled by other electrons. This electron-electron repulsion can significantly modify the effective force experienced by an individual electron.

2. **Shielding or Screening Effect**: Inner electrons can shield outer electrons from the full charge of the nucleus. As a result, outer electrons might experience a reduced effective nuclear charge, altering their energy levels and radii.

The Advent of Quantum Numbers

As scientists delved into multi-electron atoms, it

became clear that a set of quantum numbers was necessary to describe the state of an electron fully:

- **Principal Quantum Number (n)**: This number, introduced during the study of the hydrogen atom, denotes the energy level or shell of the electron.
  Larger values of n indicate electrons that are further from the nucleus and have higher energy.

- **Azimuthal Quantum Number (l)**:
  Representing the shape of the orbital, it dictates whether an electron resides in an s, p, d, or f orbital.

- **Magnetic Quantum Number (m_Y)**:
  This number describes the orientation of the orbital in space, determining its alignment relative to external magnetic fields.

- **Spin Quantum Number (m_s)**: Electrons, interestingly, have intrinsic angular momentum or 'spin'. This quantum number captures this property, having possible values of $+\frac{1}{2}$ or $-\frac{1}{2}$, often referred to as 'spin-up' and 'spin-down', respectively.

Building the Periodic Table: Aufbau Principle and Hund's Rule

With the discovery of quantum numbers, two principles became instrumental in understanding the arrangement of electrons in atoms:

- **Aufbau Principle**: Electrons occupy the lowest energy orbitals available. This 'building-up' process is methodical and explains the structure of the periodic table. Beginning with hydrogen and its single electron occupying the 1s orbital, we move to helium with two electrons in the same 1s orbital, and so on. The sequence follows 1s, 2s, 2p, 3s, and continues, with each orbital filling according to its energy level.

- **Hund's Rule**: Before any two electrons occupy an orbital in a subshell, other orbitals in the same subshell must first contain one electron each. All single electrons in a given subshell have the same spin, maximizing total electron spin.

Intricacies of Atomic Spectra in Multi-Electron Atoms

The spectral lines, which were a significant clue leading to quantum mechanics in the context of hydrogen, become far more intricate with multi-electron atoms. The splitting of spectral lines in the presence of a magnetic field, known as the Zeeman effect, or an electric field, termed the Stark effect, becomes more pronounced and complicated. These effects further reinforced the quantum nature of atoms and provided deeper insights into internal atomic structures.

Quantum Numbers and Electron Configuration

At the core of every atom, hidden beneath layers of protons, neutrons, and electron clouds, lies an intricate tapestry of quantum mechanics.

Harmony in the Atomic World: Understanding Quantum Numbers

1. **Principal Quantum Number (n)**: Reflecting the main energy level or shell of an electron, 'n' can be any positive integer. As 'n' increases, the electron's energy and its average distance from the nucleus also increase.

   For instance, electrons in the first shell (n=1) are closer to the nucleus and have lower energy compared to those in the third shell (n=3).

2. **Azimuthal Quantum Number (l)**: Dictating the shape or type of the orbital (s, p, d, f), 'l' can be any integer ranging from 0 to n-1. This number plays a pivotal role in defining the spatial distribution of an electron's probability density.

3. **Magnetic Quantum Number (m_l)**: Describing the orientation of an orbital within a subshell, 'm_l' takes on integer values from -l to +l, including zero. For a given value of 'l', there are 2l + 1 possible values of 'm_l', indicating the number of orbitals in a subshell.

4. **Spin Quantum Number (m$_s$)**: Representing the two possible spin orientations of an electron (either "spin-up" or "spin-down"), 'm$_s$' can have values of +½ or -½.

Impacts and Manifestations: The Periodic Table

The periodic table, a masterpiece of scientific organization, is deeply rooted in the principles of quantum numbers and electron configurations. Elements are ordered by increasing atomic number, but their placement in specific groups or periods reflects similarities in electron configurations.

For instance, all alkali metals, residing in the first column of the periodic table, have a single electron in their outermost s orbital. This shared trait underlies their similar chemical properties.

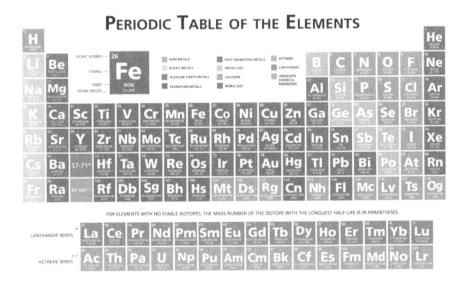

PERIODIC TABLE OF THE ELEMENTS

Chapter 8
Quantum Mechanics in Chemistry

Molecular Bonding

In the grand orchestra of molecules, individual atoms play the role of musicians. But rather than instruments, they use their electrons to produce the symphony of molecular structures we observe. Bonds between atoms are the melodies of this orchestra. But what dictates these patterns of molecular bonding? The answer, fascinatingly, lies in quantum mechanics.

The Dance of Electrons

Atoms are not solitary entities. They have a propensity to bond with other atoms, creating molecules, the building blocks of everything around us. From the air we breathe to the very cells of our bodies, molecules dominate our existence. At the heart of these molecular formations are electrons, specifically the valence electrons, which are the outermost electrons of an atom.

These valence electrons are the main players in the bonding game. When atoms approach each other, their electron clouds interact. Sometimes, they share electrons; at other times, one atom donates an electron to another.

The patterns and nature of these interactions are the essence of molecular bonding. But why do electrons behave this way? Why do some atoms share, while others donate or receive electrons? Quantum mechanics offers us answers.

Covalent Bonds

Imagine two friends sharing a secret. This sharing creates a bond between them, making their relationship stronger. In the atomic world, a similar kind of bond exists, known as a covalent bond. It's where two atoms come close enough for their valence electrons to overlap, resulting in a shared pair of electrons. This shared pair resides in a region called the molecular orbital.

But why would atoms decide to share? Delving into quantum theory, we find that when atoms share electrons, they often achieve a more stable electron configuration.

This configuration is usually similar to that of noble gases, the most stable elements on the periodic table. By sharing, atoms lower their energy, and in the universe's never-ending quest for stability, this is a favorable move.

One of the most classic examples of covalent bonding is the molecule of oxygen (O_2) that we breathe. Each oxygen atom shares two electrons with the other, creating a double covalent bond.

Ionic Bonds

While sharing is a noble act, in the atomic realm, there's also a strategy of giving and taking. Some atoms, due to their electron configurations, find it more energetically favorable to donate an electron, turning them into positive ions (cations). Others, conversely, find solace in accepting an extra electron, becoming negative ions (anions). The attraction between these oppositely charged ions results in ionic bonds.

The inception of ionic bonds is a classic representation of quantum mechanics at play. The energy levels and configurations of electrons within atoms determine whether they're more inclined to give up or accept electrons. Sodium chloride ($NaCl$), our common table salt, offers a classic example.

Sodium donates one electron to chlorine, resulting in the formation of positive sodium ions and negative chlorine ions, which then attract each other to form the crystalline structure of table salt.

Metallic Bonds

Apart from covalent and ionic bonds, there exists another type of bonding in metals called metallic bonding. In metals, atoms release their outermost electrons, which then move freely throughout the entire metal structure. This free movement creates a 'sea' of electrons surrounding positive metal ions. This sea gives metals their characteristic properties such as conductivity.

The concept of a free electron 'sea' can be understood better using quantum mechanics. Electrons in metals exist in a delocalized state. They don't belong to any specific atom but are spread out, moving freely in available energy states.

This is markedly different from the localized electrons seen in covalent or ionic bonds and highlights the versatility and complexities of quantum behaviors.

Molecular bonding, with its myriad forms and behaviors, is a testament to the power and intricacies of quantum mechanics. At a scale so minute that it's beyond human visualization, tiny particles follow principles that create the vast and diverse universe of molecules we see, touch, and experience daily.

Quantum Tunneling in Chemical Reactions

The world of quantum mechanics is rife with phenomena that defy our everyday understanding.

Just when we thought we had a grip on the subtleties of atomic and molecular interactions, quantum mechanics unveils another layer of complexity. One such marvel is quantum tunneling, a phenomenon that profoundly influences many chemical reactions.

This section will delve into the depths of quantum tunneling, its implications in chemical processes, and the pioneering scientists who have made groundbreaking discoveries in this area.

The Essence of Quantum Tunneling

To start, imagine standing in front of a massive wall, too tall to climb and too solid to break through. Conventional wisdom and our macroscopic experiences would suggest that there's no way past this barrier. However, in the quantum realm, particles like electrons can sometimes "tunnel" through barriers that should, in theory, be impenetrable.

Quantum tunneling is a direct consequence of the wave-like nature of particles.

When an electron, or any particle for that matter, approaches a barrier, its wave function can slightly extend past the barrier. If conditions are right, there's a non-zero probability that the particle can appear on the other side of the barrier, effectively tunneling through it. This phenomenon isn't just theoretical; it's been observed and validated in many experiments.

In chemical reactions, reactants have to overcome specific energy barriers to transform into products. Traditionally, we've understood this through the framework of activation energy: reactants need a certain push to cross this energy hill. However, in some reactions, especially at low temperatures, the reactants seemingly don't have enough energy. Yet, the reactions still proceed. How?

The answer lies in quantum tunneling. Even if particles don't possess the requisite activation energy, they can tunnel through the energy barrier, leading to a successful reaction. This is particularly prevalent in reactions involving hydrogen atoms, given their small mass, which enhances their wavelike properties.

A notable example is the fusion reactions within our Sun. Protons (hydrogen nuclei) collide and fuse to form helium, releasing vast amounts of energy in the process. However, the temperatures and pressures at the Sun's core, while incredibly high, are still not sufficient for these protons to overcome their mutual electrostatic repulsion (since like charges repel).

Quantum tunneling comes to the rescue, allowing these protons to come close enough to fuse.

Quantum Effects in Biological Systems

Quantum mechanics, with its wave functions, probabilities, and superpositions, was initially thought to be solely the domain of the very small – the atomic and subatomic worlds. However, the past few decades have witnessed a growing appreciation for how quantum effects might play a significant role in larger, more complex systems, notably in biology.

These quantum effects have provided potential explanations for several long-standing puzzles in biology, infusing new perspectives into our understanding of life's processes.

Quantum Coherence in Photosynthesis

Photosynthesis, the process through which plants convert sunlight into energy, is one of biology's most essential functions. At the heart of this process, in the reaction centers of plants, light photons are absorbed, creating excitons – packets of energy. These excitons need to travel to reaction centers to convert light into chemical energy efficiently.

Classical physics would predict these excitons take a random walk, similar to a drunkard's path, to find the reaction center. However, experiments, particularly those using ultrafast lasers, have indicated that the excitons don't wander randomly.

Instead, they seem to "sample" multiple pathways simultaneously and choose the most efficient one, a phenomenon reminiscent of quantum superposition.

This efficiency has been attributed to a quantum effect called "coherence," where particles become interconnected and their properties start to overlap. Coherence, in the context of photosynthesis, allows plants to transfer energy with astonishing efficiency, nearly 100% in some cases.

The exact mechanism and significance of this quantum coherence in photosynthesis are areas of active research and debate.

Quantum Tunneling in Enzyme Reactions

As previously discussed previously, quantum tunneling enables particles to cross barriers they shouldn't be able to, given classical physics rules. In biology, enzymes – proteins that catalyze chemical reactions – often deal with barriers when facilitating reactions.

It appears that several enzymes might be harnessing the power of quantum tunneling to accelerate reactions.

For instance, enzymes involved in synthesizing DNA, repairing it, or helping in respiration might utilize tunneling to transfer protons or electrons faster than classical physics would allow. The idea here is that enzymes might be evolved structures designed in part to encourage quantum tunneling, thereby making reactions more efficient.

Chapter 9
Quantum Computing

Quantum Bits (Qubits)

In classical computing, the basic unit of information is the "bit." A bit can be in one of two states: 0 or 1. It's akin to a switch that's
either off or on. However, the quantum world doesn't obey the same intuitive rules we see in our everyday life.

The fundamental unit of quantum information is the "qubit."

At its core, a qubit is a quantum system that can exist in a superposition of two states. Let's use an electron in a magnetic field as an example. The electron's spin may be in alignment with the field, representing a state $|0\rangle$, or it may be opposite to the field, representing a state $|1\rangle$. But crucially, it could also be in a superposition of these states, meaning it exists in both states simultaneously.

Qubit Representation

Graphically, qubits are often represented on a Bloch sphere. Instead of just the poles (0 and 1), any point on the sphere is a valid qubit state, with the poles representing the classical states.

A qubit can be transformed from one state to another by using a quantum gate (a topic we'll dive deeper). Quantum gates, akin to classical logic gates, allow for the manipulation of qubit states, enabling quantum computation.

One of the most profound features of qubits is their ability to be "entangled" with one another. When qubits become entangled, the state of one qubit becomes dependent on the state of another, regardless of the distance between them. This non-local property gives quantum computers a potential computational advantage in certain problems over classical computers.

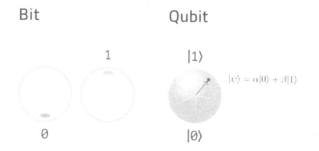

Decoherence in Quantum Computing

A significant challenge in developing quantum computers is maintaining the integrity of qubits. Due to their interaction with the environment, qubits can undergo decoherence, which destroys the valuable quantum information they hold.

Applications of Qubits

1. **Quantum Search Algorithms**: With the ability of qubits
   to exist in multiple states simultaneously, quantum computers can potentially search large databases faster than classical computers using algorithms like Grover's.

2. **Factoring Large Numbers**: A quantum computer with enough qubits could potentially factor large numbers in polynomial time using Shor's algorithm. This poses a threat to classical encryption techniques.

3. **Simulating Quantum Systems**: Quantum systems are profoundly complex. Simulating them on classical computers is often infeasible. Qubits offer a more natural representation of these systems, making simulations more tractable.

Quantum Gates and Circuits

Quantum computation, like its classical counterpart, relies on operations that transform input data to produce desired outputs. In classical computing, we use logical gates such as AND, OR, and NOT to achieve this. In the quantum realm, we have quantum gates that operate on qubits.

A quantum gate is a fundamental quantum circuit operating on a small number of qubits. They are the building blocks of quantum circuits, just as classical gates are for classical circuits. Unlike classical gates, which output a definite 0 or 1, quantum gates produce outputs that are quantum superpositions of 0 and 1.

Basic Quantum Gates

1. **Pauli Gates**:
   - **X-gate**: Analogous to the classical NOT gate, it flips a qubit from $|0\rangle$ to $|1\rangle$ and vice versa.

- **Y and Z-gates**: These introduce a phase flip to the qubit states.

2. **Hadamard Gate (H-gate)**: Often dubbed the cornerstone of quantum computation, this gate creates superposition. It transforms a qubit that's in one of its base states to a state where it has an equal probability of being measured as $|0\rangle$ or $|1\rangle$.

3. **Phase Gates**: These gates shift the phase of the $|1\rangle$ state. A common example is the S-gate, which applies a $\pi/2$ radian phase shift.

4. **Controlled Gates**: The most famous is the CNOT gate. It operates on two qubits, flipping the second qubit (the target) if the first qubit (the control) is $|1\rangle$. This gate is essential for creating qubit entanglement.

5. **Toffoli Gate**: A controlled-controlled-not gate. It operates on three qubits, flipping the third qubit only if the first two qubits are both in state $|1\rangle$.

6. **Fredkin Gate**: A controlled swap gate. It swaps the states of the second and third qubits if and only if the first qubit is in the state $|1\rangle$.

Quantum Circuitry

Much like classical circuits, quantum gates can be organized in sequences to perform more complex operations. These sequences are referred to as quantum circuits. Quantum algorithms often involve applying a series of gates to an input set of qubits and measuring the result.

However, there are crucial differences:

1. **Reversibility**: Quantum gates are reversible, meaning that for every gate there's an inverse gate that undoes its operation. This contrasts with many classical gates, which are not reversible.

2. **No Cloning Theorem**: Unlike classical bits, qubits cannot be arbitrarily copied due to the "No Cloning Theorem". This poses challenges for quantum error correction.

Quantum Parallelism

An essential feature of quantum circuits is the ability to perform many calculations at once. A set of qubits in a superposed state can be used to compute a function on many different input values simultaneously, a phenomenon called quantum parallelism.

Error Correction in Quantum Circuits

Given the fragile nature of quantum information, quantum gates are not perfect and introduce errors. Quantum error correction techniques have been developed to detect and correct these errors without observing the qubits directly. This remains one of the most active areas of quantum computing research.

Quantum Algorithms and Quantum Supremacy

Quantum algorithms leverage the principles of superposition, entanglement, and interference to solve problems that classical computers find prohibitively time-consuming or practically impossible.

Prominent Quantum Algorithms

1. **Shor's Algorithm**: Proposed by Peter Shor, this algorithm can factorize a large number into its prime constituents efficiently. It poses a direct threat to the RSA encryption widely used today, as RSA's security relies on the difficulty of factorizing large numbers. While classical computers would take millennia to factorize certain numbers, a sufficiently large quantum computer using Shor's algorithm could accomplish this in mere hours or even minutes.

2. **Grover's Algorithm**: Lov Grover's brainchild, this quantum algorithm, searches an unsorted database or solves a black-box computational problem quadratically faster than its classical counterparts. This doesn't represent an exponential speedup as in Shor's case, but it's still a considerable advantage.

3. **Quantum Fourier Transform (QFT)**: Integral to many quantum algorithms, including Shor's, the QFT is the quantum analogue of the classical discrete Fourier transform. It efficiently extracts frequency information from a quantum state and has far-reaching applications in quantum computing.

Quantum Supremacy

The term "quantum supremacy" was popularized to describe the point where quantum computers can perform tasks that classical computers practically cannot, regardless of the computational resources thrown at them. In 2019, Google claimed to have achieved quantum supremacy with its 53-qubit processor named Sycamore.

The task was specialized and of no practical use, but it was a demonstration that quantum computation had moved from theory to reality.

Challenges and Implications

Despite the promising advancements, quantum algorithms face several challenges:

1. **Error Rates**: Quantum computers are highly susceptible to errors. Even minor external disturbances can cause qubits to change their state (decoherence). Therefore, running complex algorithms that require thousands of gates can produce unreliable outcomes.

2. **Algorithm Complexity**: Designing quantum algorithms is inherently more challenging than classical algorithms. Often, translating a quantum advantage in theory to practical use cases becomes intricate.

3. **Hardware Limitations**: Current quantum computers are in their infancy, often termed "Noisy Intermediate-Scale Quantum" (NISQ) devices. Their limitations restrict the execution of more complex algorithms.

Pioneers and Their Contributions

While we've mentioned Peter Shor and Lov Grover, many others have contributed profoundly to this field. For instance, Seth Lloyd proposed the first quantum algorithms for the simulation of physical systems. Barbara Terhal and John Preskill have worked extensively on quantum error correction, a crucial aspect for running fault-tolerant quantum algorithms.

Chapter 10
Quantum Field Theory

Quantum Electrodynamics (QED)

Quantum Field Theory (QFT) provides a robust lens to perceive and understand the nuanced dynamics of particle interactions at extremely minute levels, and among its illustrious branches, Quantum Electrodynamics (QED) stands out, serving as a beacon for other quantum theories. QED intricately delves into the interactions between light, quantified as photons, and matter, with a predominant focus on electrons and their antiparticle counterparts, positrons.

At the heart of QED lies the electromagnetic force, a linchpin among the universe's quartet of foundational forces. From an elementary perspective, this force functions straightforwardly: like charges repel each other, while opposite charges exhibit

attraction. However, this seemingly simple principle evolves into a layered and complex dance when zoomed into the quantum realm. It's in this microscopic world that QED articulates how charged particles swap photons to mediate these forces.

QED is not merely a theory but a tale of human ingenuity. The story can't be narrated without emphasizing the paramount contributions of physicist Richard Feynman. His ingenuity gave birth to Feynman diagrams, a graphical method to represent particle interactions. These diagrams weren't mere pedagogic tools but encapsulated deep mathematical truths. For instance, a rudimentary Feynman diagram might depict an electron and a positron meeting and annihilating each other, resulting in the birth of a photon.

This photon, a transient entity, might then decay, birthing yet another electron-positron pair.

The raison d'être for QED emerges from the shortcomings of classical theories when addressing minuscule scales. While classical electromagnetism, bolstered by legends like James Clerk Maxwell, proficiently expounded vast, macroscopic phenomena, it fumbled when probing the depths of atomic and subatomic realms. Conspicuous disparities like the Lamb shift—a minute shift in the energy levels of an electron in a hydrogen atom—and the anomaly in the electron's magnetic moment stood as unsolved enigmas. With QED's inception, these puzzles found their answers. Merging the inherent probabilistic nature of quantum mechanics with the rigorous light-speed limitations set by relativity, QED presented a comprehensive solution set.

Yet, the adoption of QED wasn't instantaneous. It demanded experimental validations, theoretical refinements, and the relentless passion of a community of physicists. Among these, Julian Schwinger and Sin-Itiro Tomonaga charted paths parallel to Feynman. Their collective endeavors ushered QED to its contemporary stature.

Their groundbreaking work didn't go unnoticed, with the Nobel Prize in Physics in 1965 commemorating their contributions. This recognition not only elevated their individual legacies but also firmly entrenched QED as the cornerstone for successive quantum field theories.

The implications of QED are vast. Beyond just providing a mathematical framework, it expanded our horizons, enabling us to fathom the intricacies of the universe. It bridged the macroscopic world we experience daily with the enigmatic quantum world, affirming that the laws governing minute particles also sculpt the vast cosmic tapestry.

QED's journey, from its nascent stages to its monumental achievements, serves as a testament to humanity's relentless quest for knowledge.

Quantum Chromodynamics (QCD)

Diving further into the enigmatic waters of Quantum Field Theory (QFT), we approach a compelling chapter, Quantum Chromodynamics (QCD). If QED handles the intimate tales of photons, electrons, and their relatives, QCD narrates the epic of the strong nuclear force and the fascinating particles it governs: quarks and gluons.

To navigate the quantum landscape, one first needs to understand the structure of atomic nuclei. Enclosed within this small space are protons and neutrons, but these are not elementary. Instead, they're a bustling conglomerate of quarks, bound together by the exchange of gluons. While QED is the theory of electromagnetic interactions, QCD is the quantum theory dedicated to explaining the interactions mediated by the strong force.

The name "chromodynamics" is a nod to the use of the term "color," albeit metaphorically. It has nothing to do with the colors of the visual spectrum but is a novel way of differentiating quarks. Just as electric charge is the source of electromagnetic interactions, "color charge" is the source of strong force interactions. Quarks come in three "colors" - red, blue, and green, while gluons are carriers of these color charges.

This choice of nomenclature becomes particularly poignant when one understands that just like colors in the visible spectrum combine to give white light, quarks combine in a manner where their colors "neutralize" each other, ensuring that observable matter is always "white" or color-neutral.

QCD is deeply fascinating, especially in how it deviates from QED in behavior. When quarks come closer, the force binding them decreases, a phenomenon termed "asymptotic freedom." This concept, albeit counterintuitive compared to our everyday experiences, was a revolutionary discovery. It showed that at extremely high energies or short distances, quarks behaved almost like free particles.

Conversely, as quarks attempt to distance themselves, the force between them intensifies, forever imprisoning them within protons, neutrons, and other hadrons.

In the mid-20th century, when particle accelerators began unveiling a zoo of particles, the need for an organizing principle became paramount. The discovery of quarks by Murray Gell-Mann and George Zweig in the 1960s provided the first hint. Yet, understanding how these quarks interacted and the forces binding them inside hadrons needed a more sophisticated theory. This quest culminated in the birth of QCD in the 1970s.

The experimental and theoretical underpinnings that shaped QCD are vast. Deep inelastic scattering experiments, which bombarded protons with high-energy electrons, unveiled the innards of protons and provided the first evidence of quarks. Later, the observation of particle jets corresponding to quarks and gluons in high-energy experiments added another feather in QCD's cap.

Theoretical advancements were no less monumental. Pioneers like David Gross, Frank Wilczek, and David Politzer, who played pivotal roles in understanding the strong force's behavior at short distances, were recognized with a Nobel Prize in 2004 for their foundational work on asymptotic freedom.

One of the grand successes of QCD was its incorporation into the Standard Model of particle physics. This model, which remains our best description of the subatomic world, classifies all known elementary particles and three of the four fundamental forces (excluding only gravity). The seamless integration of QCD into this model solidified its place in the pantheon of successful quantum theories.

QCD is not merely a set of equations but a bridge to understanding the universe's essence. From the cores of neutron stars to the first moments after the Big Bang, understanding the dynamics of quarks and gluons is indispensable.

The Higgs Field and the Standard Model

The world of particle physics, prior to the 1960s, was in turmoil. While the Standard Model was being developed as a coherent framework to describe the elementary particles and their interactions, one crucial puzzle remained unsolved:

Why do particles have mass? In the elegant mathematical formulations that described particles and their interactions, the W and Z bosons (mediators of the weak force) were predicted to be massless, but this was in stark contrast to experimental observations.

Enter the concept of spontaneous symmetry breaking and the Brout-Englert-Higgs mechanism. Theorists, including Peter Higgs, François Englert, and Robert Brout, proposed an omnipresent field permeating all of space: the Higgs Field. Particles, when they move through this field, interact with it. Depending on the strength of this interaction, some particles acquire mass while others remain massless. Think of the

Higgs Field as a viscous medium, like honey. While some objects (like small pebbles) might pass through honey with relative ease, larger objects (like a spoon) would face resistance. This resistance, in the quantum world, is what we interpret as mass.

The Higgs boson itself is a manifestation or ripple in the Higgs Field. Just as photons are quanta of the electromagnetic field, the Higgs boson is a quantum of the Higgs Field. Theoretical physicists were thrilled with this elegant solution, but without empirical evidence, this remained a mere mathematical curio.

The quest for the elusive Higgs boson spanned decades and required the most advanced particle accelerators. The Large Hadron Collider (LHC) at CERN, Geneva, became the focal point of this hunt. Its massive tunnels, sprawling over 27 kilometers, accelerated particles to nearly the speed of light, smashing them together and scouring the resultant debris for the Higgs boson.

Beyond the Standard Model, the Higgs boson's discovery raises profound questions and opens new avenues of research. It begs the question: What else don't we know? What lies beyond the Higgs? How does the Higgs mechanism interact with the mysteries of dark matter and dark energy? While the Higgs discovery filled a significant gap in our understanding, it, in many ways, is the beginning of a new chapter in the tale of quantum physics. The cosmos remains vast, enigmatic, and ever-inviting, with many more secrets awaiting to be unraveled.

Chapter 11

Quantum Gravity and the Search for a Unified Theory

General Relativity and Quantum Mechanics

At the dawn of the 20th century, two radical theories emerged, reshaping our understanding of the universe. On one hand, Albert Einstein's theory of General Relativity painted a picture of a universe where massive objects warp the very fabric of spacetime, leading to the gravitational attraction we observe. On the other, Quantum Mechanics emerged as a theory to describe the behavior of the tiniest particles in the universe, at scales where classical physics breaks down.

General Relativity

Einstein's theory of General Relativity, presented in 1915, is a theory of gravitation. It proposes a universe where spacetime isn't a static stage on which events unfold, but rather, a dynamic entity influenced by matter and energy. Massive objects, like stars and planets, distort this spacetime fabric around them. Imagine a stretched-out trampoline.

If you were to place a heavy ball in its center, it would create a depression. Similarly, Earth warps the spacetime around it, and this curvature tells smaller objects, like the Moon or satellites, how to move.

General Relativity has passed numerous experimental tests, from the bending of light around massive objects (gravitational lensing) to the recently detected gravitational waves, ripples in spacetime caused by violent cosmic events.

Quantum Mechanics

If General Relativity deals with the massive, Quantum Mechanics handles the minute. It delves into the probabilistic nature of particles at atomic and subatomic scales. The theory was born out of necessity when classical physics couldn't explain certain phenomena, such as the spectral lines of hydrogen or the behavior of electrons in metals.

Quantum Mechanics introduces concepts like superposition, where particles can exist in multiple states simultaneously, and wave-particle duality, where particles exhibit both wave-like and particle-like characteristics depending on how we observe them.

The Clash of Titans

While both General Relativity and Quantum Mechanics are incredibly successful in their respective domains, they fundamentally clash in their descriptions of the universe. For instance, General Relativity's smooth spacetime continuum contrasts with Quantum Mechanics' discrete, quantized nature.

The most famous battleground for this clash is black holes. At the heart of a black hole lies the singularity, a point of infinite density.

Here, the gravitational forces (described by General Relativity) become so strong that Quantum Mechanical effects can't be ignored. Yet, our current understanding can't reconcile the two. Another realm of conflict is the early universe, just moments after the Big Bang, where quantum fluctuations might have influenced the structure of the entire universe.

Prominent physicists like Stephen Hawking and Roger Penrose have made significant strides towards harmonizing these two theories. Hawking, for example, proposed that black holes could emit radiation due to quantum effects near the event horizon, which became known as "Hawking Radiation". Yet, despite these efforts, a full theory of "Quantum Gravity" remains one of the biggest unsolved puzzles in physics.

String Theory and Loop Quantum Gravity

One of the grand challenges of modern physics is unifying the cosmos's vast expanse with the tiny, tumultuous world of quanta. Both General Relativity and Quantum Mechanics have been incredibly successful within their realms of applicability.

However, when they intersect, especially in extreme conditions like black holes' centers or the universe's birth, they clash vehemently. Two leading approaches, among others, have been suggested to address this discord: String Theory and Loop Quantum Gravity.

String Theory: A Symphony of Vibrations

The essence of String Theory is captivatingly simple and beautiful: instead of viewing the smallest constituents of matter as particles, they are one-

dimensional "strings" that vibrate at different frequencies. These vibrations correspond to various particles. For instance, an electron isn't a point-like particle but a tiny string vibrating in a specific pattern. The inception of String Theory came in the late 1960s, inspired by attempts to explain the strong nuclear force. By the 1980s, it was seen as a potential 'Theory of Everything', encompassing all fundamental forces.
Key Aspects of String Theory:

1. **Higher Dimensions**: String Theory necessitates more than the familiar three spatial dimensions and one time dimension. Specifically, the theory requires a total of 10 or 11 dimensions, depending on the version. These extra dimensions might be compactified or hidden from our direct perception, shaping the universe's properties in ways we're just beginning to fathom.

2. **Multiple Versions**: There isn't one single String Theory. Five consistent versions were developed, which led to the concept of M-theory, proposed in the 1990s, as a potential unifying framework.

3. **Branes**: In addition to strings, the theory introduces "branes" - multidimensional objects that can range from 0-dimensional (point-like) to 9-dimensional (spreading throughout the entire universe).

Loop Quantum Gravity: Quantizing Spacetime

Where String Theory ventures into higher dimensions, Loop Quantum Gravity (LQG) takes a different path. It remains within our four-dimensional spacetime but seeks to quantize it.

Fundamentals of LQG:

1. **Quantized Spacetime**: In LQG, spacetime isn't continuous. It's made up of tiny, discrete loops or "spin networks". These loops represent quantized units of area and volume, suggesting a minimum possible distance in the universe, below which the concept of space doesn't hold.

2. **No Need for Extra Dimensions**: Unlike String Theory, LQG doesn't require additional dimensions. It operates within the familiar four-dimensional spacetime.

3. **Black Hole Entropy**: One of the significant successes of LQG is the calculation of black hole entropy, which aligns well with predictions made using other methods.

The Problem of Time in Quantum Gravity

The fusion of quantum mechanics and general relativity remains one of the most tantalizing and formidable challenges in theoretical physics. As these two titanic theories converge, a central issue that emerges is the problem of time. It's not merely a philosophical inquiry but a critical stumbling block, with implications that reverberate throughout the entire effort to discover a theory of quantum gravity.

The Classical Perspective on Time

Traditionally, in the Newtonian view, time is absolute. It flows uniformly, serving as a constant backdrop against which all physical events play out. This clockwork concept of time is rigid and unwavering, and every observer, regardless of their state of motion, would agree on the sequence and duration of events.

General relativity, Einstein's magnum opus, dramatically altered this perspective. In his spacetime continuum, time and space intermingle, bending and warping in response to mass and energy. The stronger the gravitational field, the more pronounced the warping. This leads to time dilation, where time can flow differently depending on one's relative motion or gravitational environment.

Quantum Mechanics and the Notion of Time

In the quantum realm, time retains its Newtonian flavor: it's absolute. The Schrödinger equation, which dictates the evolution of quantum systems, is set against this absolute time. There's no 'quantum time dilation' or similar phenomenon. For all the strange behaviors of quantum particles, their dance is choreographed on the unwavering stage of constant temporal progression.

The Conundrum: Reconciling Two Times

Marrying these two perspectives leads to a conundrum. How can time be both malleable, as general relativity suggests, and unyielding, as quantum mechanics requires? This "problem of time" isn't just academic. It fundamentally impedes our efforts to describe things like the early universe's state or the interiors of
black holes, where both quantum effects and intense gravitational fields play roles.
Some approaches to this issue include:

1. **Canonical Quantum Gravity**: This technique attempts to quantize gravity as if it were any other force. However, it leads to a version of the Schrödinger equation that's filled with infinities and doesn't yield a clear concept of time.

2. **The "Timeless" Approach**: Some theories, like the Wheeler-DeWitt equation, propose a universe without time. In these views, what we perceive as time is just an emergent phenomenon, resulting from our particular perspective within a larger, timeless, block universe.

3. **Relational Quantum Mechanics**: Proposed by the likes of Carlo Rovelli, this perspective posits that events are only meaningful in relation to others. Time, thus, isn't an absolute backdrop but arises from the relationships between quantum systems.

Chapter 12

Interpretations of Quantum Mechanics

The Copenhagen Interpretation

In the ever-evolving landscape of quantum mechanics, multiple interpretations have arisen to explain the seemingly paradoxical behaviors of particles at the quantum scale. Each interpretation attempts to provide a framework or lens through which we can understand the results of quantum experiments and the nature of reality itself. Among these, the Copenhagen Interpretation stands as one of the most historically influential and extensively taught.

Historical Context and Development

The birth of the Copenhagen Interpretation traces back to the 1920s, predominantly in the city from which it gets its name - Copenhagen, Denmark. Spearheaded by Niels Bohr and Werner Heisenberg, this interpretation was among the first to attempt a conceptual scaffold for the newly formulated quantum theory.

Bohr and Heisenberg, along with other physicists like Max Born, were grappling with the counter-intuitive experimental results that defied classical intuition. The double-slit experiment, for instance, showcased particles behaving both as waves and particles, depending on whether they were observed. This duality raised profound questions about the role of the observer in experimental outcomes.

Core Tenets of the Copenhagen Interpretation

1. **Wave Function and Probability**: Central to the Copenhagen Interpretation is the concept of the wave function, denoted often as Ψ. Rather than depicting a real physical entity, the wave function encapsulates the probabilities of finding a quantum system in a particular state. Once a measurement is made, the wave function "collapses" to a singular state, and it is this act of measurement that brings about a definitive outcome.

Example: Consider an electron in a potential well. Prior to measurement, the electron doesn't have a definitive position. When measured, the wave function collapses, and the electron is found in a specific location within the well.

2. **Observer Effect**: The act of observation plays a paramount role. A quantum system remains in a superposition of states until a measurement is made, at

which point it collapses to a definitive state. The very act of observation influences the state of the system.

Example: In the double-slit experiment, electrons sent towards a barrier with two slits display an interference pattern, suggesting wave-like behavior. However, if one tries to "watch" which slit the electron passes through, the interference disappears, implying particle-like behavior.

3. **Classical-Quantum Boundary**:
   The Copenhagen Interpretation postulates a boundary between the quantum world and the classical world. While quantum effects dominate at microscopic scales, classical physics takes over at macroscopic scales.

Example: Quantum effects might dominate in the behavior of an individual electron around an atomic nucleus, but when considering a tennis ball being thrown, classical mechanics provides a sufficient and accurate description.

The Many-Worlds Interpretation

As quantum mechanics grew and developed, so did the discomfort with some of its foundational concepts. This unease, particularly around the idea of wave function collapse and the central role of an observer in determining the state of a quantum system, led to alternative interpretations. Among the most daring and provocative of these is the Many-Worlds Interpretation (MWI). In contrast to the Copenhagen Interpretation's subjective randomness, the MWI posits a universe where every quantum possibility is realized in a branching, deterministic multiverse.

Origins and Philosophical Grounding

The roots of the Many-Worlds Interpretation can be traced back to Hugh Everett III's 1957 doctoral thesis,

"The Theory of the Universal Wave Function." Disturbed by the perceived inadequacies and ambiguities in the Copenhagen Interpretation, Everett sought a more objective description of quantum mechanics, free from the ambiguities surrounding the observer and the wave function collapse.

Core Principles of the Many-Worlds Interpretation

1. **Universal Wave Function**: At the heart of MWI is the belief in a universal wave function that evolves deterministically over time. There's no "collapse" as such. Every potential outcome of a quantum measurement already exists in this universal wave function.

Example: If an atom might decay or not decay, both possibilities coexist in the wave function. Upon measuring, we simply find ourselves in one branch of reality or another.

2. **Branching Realities**: Every quantum event that has multiple possible outcomes leads to a branching of the universe. Each potential outcome occurs in its own separate "branch" or "world." This results in an ever-expanding multiverse of non-communicating parallel realities.

Example: Consider the famous thought experiment of Schrödinger's cat, where a cat is both alive and dead inside a box until observed. In the MWI, both outcomes occur but in separate branches of reality.

In one branch, you open the box to find a live cat, and in another parallel universe, you discover the cat is dead.

3. **No Special Role for Observers**: In stark contrast to the Copenhagen Interpretation, observers don't have any special status in the MWI. Observers, like everything else, are part of the quantum system and, thus, subject to its branching realities.

Example: When measuring a quantum system, an observer becomes entangled with it, leading to multiple copies of the observer existing across different branches, each having perceived a different outcome.

Implications and Debates

The Many-Worlds Interpretation, while resolving some of the perceived shortcomings of the Copenhagen Interpretation, brings with it a host of new challenges and debates.

1. **The Nature of Reality**: MWI challenges our most foundational notions of reality. If countless branches of reality are constantly emerging, then our sense of a singular, shared reality becomes profoundly disrupted.

2. **Probability**: One of the criticisms of MWI revolves around the interpretation of quantum probabilities. If every outcome happens in some branch, how do we make sense of probabilities in quantum mechanics?

3. **Occam's Razor**: Some critics argue that the MWI is not parsimonious. The idea that countless universes emerge with every quantum event might seem like an over-complication, defying the principle of Occam's Razor, which states that one should not make more assumptions than necessary.

Quantum Bayesianism and Other Interpretations

Venturing beyond the familiar terrains of the Copenhagen and Many-Worlds interpretations, a plethora of alternative interpretations attempt to decode the enigma of quantum mechanics. One of the most intriguing among these is Quantum Bayesianism, or QBism for short, which beckons us to consider the role of personal belief in quantum theory. Alongside QBism, a myriad of other interpretations offer unique perspectives that enrich our understanding and continue to spark debate.

Quantum Bayesianism (QBism)

At its core, QBism stands apart by viewing the wave function not as an objective description of reality, but rather as a subjective representation of an observer's beliefs or degrees of belief about a quantum system. It suggests that quantum states are essentially expressions of an observer's personal degrees of
belief about the outcomes of future experiments. This perspective moves away from a universal wave function, placing emphasis on individual observers.

Contrary to classical Bayesianism, which deals with probabilities in a deterministic universe, QBism deals with probabilities in the fundamentally uncertain quantum realm. Here, the process of updating beliefs upon receiving new information becomes vital.

When an observer makes a measurement in QBism, the wave function's update does not represent a physical change in the world, but rather a change in the observer's beliefs about the world.

To many, the subjective stance of QBism is both its strength and its point of contention. It offers a way out of many quantum paradoxes by positing that there is no objective quantum state.

Chapter 13

Quantum Phenomena in the Macroscopic World

Quantum Effects in Electronics

The dawn of the 20th century was marked by a series of groundbreaking discoveries that redefined our understanding of the universe. One of the most significant of these was the formulation of quantum mechanics, which described the strange and counterintuitive behaviors observed at the smallest scales of existence.

As perplexing as the principles of quantum mechanics were, they presented a new frontier for exploration, innovation, and application. And as we embarked on this journey into the quantum realm, it soon became evident that the technologies of the future would be deeply influenced, if not entirely shaped, by these newfound principles.

Electronics, the branch of physics and technology concerned with the design of circuits using transistors and microchips, is a testament to the profound implications of quantum mechanics in our daily lives.

As devices became smaller and more efficient, the classical models, which had successfully powered the first wave of electronic innovation, began to falter. Quantum effects, once the exclusive domain of high-end laboratories and research institutions, started playing a crucial role in the everyday function of these devices.

The Birth of Quantum Electronics

The origins of quantum electronics can be traced back to the 1920s and 1930s. Scientists and researchers of the era began noticing anomalies in the behavior of electronic devices as they approached smaller scales. One such scientist, Paul Dirac, predicted the existence of anti-electrons or positrons, effectively introducing the concept of quantum electron states. This was a monumental leap, suggesting that electrons could exist in discrete states, each with its own unique quantum properties.

The Quantum Tunneling Phenomenon

One of the most groundbreaking discoveries in quantum electronics was the phenomenon of quantum tunneling. In classical physics, an electron moving toward a barrier would either pass through if it had enough energy or be reflected back if it didn't. However, in the quantum realm, there was a non-zero probability that the electron could "tunnel" through the barrier, even if its energy was insufficient by classical standards.

This phenomenon was utilized in the development of the tunnel diode, an essential component in many of today's electronic devices. The tunnel diode showcased that, under specific conditions, electrons could move faster than what classical physics predicted, allowing for ultra-high-speed electronic functions.

Real-World Implications: From Computers to Communication

The influence of quantum effects in electronics isn't limited to specialized devices. Modern computers, smartphones, and almost all electronic gadgets we use daily rely on quantum principles to function.

For instance, the transistors in computer chips, which have shrunk to incredibly tiny scales, operate in a realm where quantum effects are not just significant but dominant. The very speed and efficiency of our devices are owed, in large part, to quantum mechanics.

Similarly, the field of communication has benefited immensely. Techniques rooted in quantum mechanics, such as quantum key distribution, promise to revolutionize the way we secure our communications, making them nearly impervious to eavesdropping.

As we continue to push the boundaries of what's possible in electronics, the principles of quantum mechanics will play an even more significant role. From ensuring the continued miniaturization of electronic components to pioneering entirely new technologies like quantum computers, the quantum realm holds the key. The journey, which began with the mere observation of anomalies at the microscale, has evolved into a full-fledged discipline, driving innovation and redefining the future of technology.

Quantum Phenomena in Optics

Optics, the study of light and its interactions with various materials, has been an essential field in physics for centuries. Yet, with the evolution of quantum mechanics in the 20th century, a fascinating intersection emerged between the age-old study of light and the new, counterintuitive theories of the quantum world.

This confluence led to the exploration of quantum phenomena within the realm of optics, presenting groundbreaking insights and potential technological advances.

Quantum Entanglement in Optics

Entanglement is a distinctly quantum phenomenon where particles become intertwined in such a way that the state of one instantaneously influences the state of the other, regardless of the distance separating them. In optics, this phenomenon is manifest in the entanglement of photons.

Such entangled photon pairs, often produced in processes like spontaneous parametric down-conversion, have been instrumental in experiments testing the fundamentals of quantum mechanics and in applications like quantum cryptography.

Photon Antibunching

One of the more subtle yet profound outcomes of quantum optics is the phenomenon of photon antibunching. Unlike classical light sources, which can emit multiple photons simultaneously, certain quantum light sources, such as single-atom emitters, emit photons one at a time.

This non-coincidental emission, where the chances of two photons being detected at the same instant are minimal, serves as a clear testament to the quantized nature of light.

Quantum Metrology

Harnessing the unique properties of quantum states, quantum metrology aims to make measurements with precision surpassing classical limits. Using states like squeezed light, where quantum uncertainties in specific properties of the light wave are reduced, quantum metrology can achieve unprecedented accuracies in measuring quantities like time, phase, and frequency.

Applications in Quantum Information and Computing

Optics has paved the way for practical quantum technologies. Quantum key distribution, an encryption scheme that's theoretically unbreakable, relies on the principles of quantum optics.

Moreover, optical lattices and trapped ion systems, both crucial for quantum computing, are primarily governed by interactions of light with matter on a quantum level.

The exploration of quantum phenomena in optics offers a harmonious blend of foundational physics and cutting-edge technology. As we venture deeper into understanding light in the quantum realm, we not only unravel the mysteries of the universe but also harness these insights to shape the next generation of technological marvels.

Whether it's in ensuring secure communications or developing lightning-fast computers, the quantum dance of photons is set to illuminate our path forward.

Quantum Effects in Biological Systems

The burgeoning field of quantum biology offers an exciting juncture between the enigmatic world of quantum mechanics and the vast complexities of biological phenomena.

This integration has led researchers to reconsider how some fundamental processes in biology might be influenced or even driven by quantum principles.

Quantum Coherence in Photosynthesis

Photosynthesis is undeniably one of the most critical processes for life on Earth, acting as the primary energy source for the planet's biosphere.

At its core, photosynthesis involves converting light energy into chemical energy stored in the form of glucose and other organic molecules.

Researchers, by delving into the intricate details of this process, have uncovered startling evidence suggesting quantum mechanics plays a role.

Enzyme Function and Quantum Tunneling

Enzymes are nature's catalysts, responsible for accelerating chemical reactions that sustain life. Classical biology teaches us about the lock-and-key model, where substrates fit into enzymes just as a key fits into a lock. However, the nuances of how enzymes achieve such astonishing speed in reactions remained a puzzle until quantum mechanics entered the scene.

Studies have indicated that some enzymes utilize quantum tunneling to facilitate reactions. In a classical world, particles need to overcome energy barriers for reactions to occur.

But in the quantum realm, particles like electrons can bypass these barriers entirely through a process called tunneling.

This essentially means that electrons can 'disappear' from one side of a barrier and 'reappear' on the other without ever crossing it.

Avian Navigation and Quantum Entanglement

Migratory birds have long captivated scientists and laymen alike with their remarkable navigational abilities.

The European robin, for instance, embarks on vast migratory treks, seemingly having an innate compass.

Research has now proposed a theory that's nothing short of fascinating: birds might be using quantum entanglement, one of the most mysterious quantum phenomena, to navigate.

When photons from sunlight interact with certain molecules in the bird's retina, they might produce pairs of radicals (molecules with unpaired electrons) that are quantum entangled.

This means that the state of one radical can instantaneously affect the state of the other, no matter the distance between them. These entangled radicals could be sensitive to the Earth's

magnetic field, providing the bird with quantum information about its orientation.

Chapter 14

Quantum Technologies of the Future

Quantum Cryptography and Communication

The advent of quantum mechanics in the 20th century was nothing short of revolutionary. It not only reshaped our fundamental understanding of nature at the most microscopic scales but also paved the way for a new frontier of technology and innovation.

One of the most tantalizing and promising applications of quantum principles lies in the realm of information science - specifically, in the fields of cryptography and communication.

The Need for Quantum Cryptography

Cryptography, the science of secret communication, is a discipline as old as civilization itself. From ancient ciphers used by monarchs to hide their messages to modern-day encryption algorithms that secure our online transactions, the need to communicate securely has always been paramount.

In our digital age, cryptography has taken center stage in ensuring the confidentiality and integrity of data transmission. However, as computational capabilities grow, many traditional cryptographic methods are under threat from the potential of quantum computers. These immensely powerful machines, once fully realized, could break widely-used encryption protocols in a matter of seconds.

This looming threat necessitates a new paradigm of encryption, and that's where quantum cryptography steps in.

Quantum Key Distribution

Central to quantum cryptography is the concept of Quantum Key Distribution (QKD). Instead of relying on mathematical complexity as traditional cryptographic methods do, QKD uses the fundamental principles of quantum mechanics to ensure security.

Imagine sending a secret message using particles of light (photons). Due to the inherent properties of quantum mechanics, any attempt to intercept and measure these photons would disturb their quantum state. This disturbance can be detected by the sender and receiver, alerting them to the presence of an eavesdropper.

Thus, eavesdropping becomes not just computationally challenging, but physically impossible without detection, providing a level of security that's unattainable through classical means.

Quantum Communication and the Future

Beyond encryption, the principles of quantum mechanics have also been harnessed for quantum communication. Researchers have been able to 'teleport' information from one location to another using quantum entanglement, a phenomenon where particles become interconnected and the state of one particle instantly influences the state of another, regardless of the distance separating them.

Quantum communication has the potential to revolutionize the way we transmit data, ensuring not only unparalleled security but also speeds and efficiencies beyond what's possible with classical systems.

Several pioneering experiments have been conducted globally to test and improve quantum communication infrastructure. Countries like China have already launched quantum satellites that aim to establish ultra-secure quantum communication networks in space.

The Road Ahead

While quantum cryptography and communication promise a future of ultra-secure and efficient data transmission, challenges remain. Maintaining the delicate quantum states over long distances, ensuring practicality for everyday use, and integrating quantum systems with existing infrastructure are all hurdles that researchers are fervently trying to overcome.

Prominent figures in the world of quantum research, such as Charles H. Bennett and Gilles Brassard, who first introduced the concept of QKD, and Anton Zeilinger, known for his work in quantum teleportation, are at the forefront of these endeavors. Their contributions, along with those of countless others, are pushing the boundaries of what's possible, ensuring that the quantum revolution permeates every facet of our digital age.

In conclusion, the realm of quantum cryptography and communication is still in its nascent stages, but its potential is vast. As we advance in our understanding and practical application of quantum principles, we stand on the cusp of a new era where data transmission is not only faster and more efficient but also impervious to threats, ensuring a safer, more secure digital future for all.

Quantum Sensing and Imaging

In a world dominated by classical physics, quantum mechanics unveils a reality filled with superpositions, entanglements, and the fundamental probabilistic nature of the universe.

These phenomena, initially seen as mere curiosities or philosophical conundrums, are now at the forefront of technological advancements, particularly in the realm of sensing and imaging.

Quantum sensing and imaging capitalize on the unique properties of quantum systems to achieve measurements and visual resolutions previously thought impossible.

The Quantum Advantage

At the heart of quantum sensing and imaging is the concept of using quantum systems, like atoms, ions, or photons, to make precise measurements. One advantage quantum systems offer is their extreme sensitivity to external influences. For instance, when quantum states of particles are disturbed, these disturbances can be accurately detected, allowing for more precise measurements than classical systems.

Quantum Metrology: Measuring with Precision

Quantum metrology revolves around the idea of using quantum principles to enhance measurement precision. Techniques based on entanglement and superposition are deployed to achieve results surpassing classical limits. For example, atomic clocks, which use the vibrations of atoms to keep time, are a product of quantum metrology. The atomic clock's extreme precision has revolutionized timekeeping and, by extension, global positioning systems (GPS) that rely heavily on synchronicity and accuracy.

Quantum Imaging: Seeing Beyond Classical Limits

Quantum imaging is another domain that exploits the weirdness of quantum mechanics to achieve unparalleled resolutions. One method, quantum ghost imaging, employs pairs of entangled photons.

While one photon interacts with the object being imaged, its entangled partner, which has never interacted with the object, can be used to create the image. This process, which seems almost magical, can achieve imaging results in conditions where classical light sources would fail.

Another significant achievement in quantum imaging is the development of cameras sensitive enough to detect single photons. This sensitivity provides the ability to produce images with minimal light, paving the way for new medical imaging techniques or surveillance in near-total darkness.

Quantum Sensors: Delving Deeper into the Unknown

Exploiting quantum phenomena such as superposition and entanglement, quantum sensors offer sensitivities unattainable by their classical counterparts.

Quantum sensors have potential applications across diverse fields, from detecting gravitational waves in astrophysics to observing minute biological processes within cells.

One notable example is the use of NV (nitrogen vacancy) centers in diamonds for sensing. These quantum systems can detect changes in temperature, electromagnetic fields, and even the presence of other quantum systems with unparalleled precision.

Interdisciplinary Applications

Both quantum sensing and imaging aren't confined to pure physics. Their implications stretch across various disciplines:

- **Medicine:** Quantum imaging can lead to more refined imaging techniques, enabling doctors to spot smaller tumors or irregularities within the body with lesser radiation exposure.
- **Astronomy:** Quantum sensors could vastly improve telescopes, allowing astronomers to detect fainter signals, leading to a better understanding of the cosmos.
- **Environmental Monitoring:** Quantum sensors can detect minute changes in environmental conditions, from water quality to atmospheric changes, aiding in timely interventions.

Quantum Materials and Quantum Metamaterials

In the continually evolving landscape of quantum mechanics, the discovery and exploration of quantum materials and quantum metamaterials have garnered significant attention. These materials, characterized by their distinct quantum properties, promise revolutionary applications, from superconductivity to radically new electronics.

Quantum Materials: A Primer

Quantum materials are substances that exhibit exotic physical properties directly resultant from quantum mechanical effects. Unlike everyday materials where quantum effects are statistically averaged out and thus not directly observable, in quantum materials, these effects manifest at macroscopic scales. This manifestation leads to an array of phenomena, many of which remain subjects of intense research and exploration.

A classic example of a quantum material is the high-temperature superconductor.

Traditional superconductors operate at very low temperatures, close to absolute zero.

However, high-temperature superconductors can exhibit zero electrical resistance at higher, albeit still cold, temperatures.

Their behavior, while not fully understood, is a direct consequence of quantum mechanics.

Quantum Metamaterials: Beyond Natural Properties

While quantum materials derive from naturally occurring quantum effects, quantum metamaterials are engineered structures designed to manipulate and control quantum phenomena in specific ways. These metamaterials can exhibit properties not found in nature, largely because of their intricate and artificially designed architectures.

Prominent Quantum Materials

Beyond high-temperature superconductors, various quantum materials have drawn scientific interest:

- **Topological Insulators:** These are materials that act as insulators in their interior but conduct electricity on their surface. What's fascinating is that this surface conductivity is immune to most types of disturbances, making them prime candidates for robust electronic devices and potentially quantum computing applications.

- **Quantum Spin Liquids:** These are materials where electron spins (typically visualized as tiny bar magnets) do not order in any regular pattern, even at very low temperatures. Instead, they form a 'liquid-like' state of constant fluctuations. This state might hold clues to high-temperature superconductivity and could have applications in quantum computing.

Chapter 15

Conclusion

The journey of quantum mechanics has been nothing short of mesmerizing. Over the past century, from the pivotal works of Einstein, Planck, and Bohr to modern quantum experiments and theories, the quantum world has reshaped our understanding of the universe's fabric. But like any great tale of exploration, while many secrets have been unveiled, countless others remain shrouded in mystery.

The realm of quantum physics, despite its immense advances, is still an unfinished canvas, with enigmas and challenges that continue to captivate and perplex the brightest minds.

One cannot help but marvel at the progress made. Quantum mechanics has successfully explained phenomena from the minute scales of subatomic particles to the expansive realms of black holes and cosmology.

Its principles have been tested, verified, and utilized in myriad ways, leading to technological revolutions and deep philosophical reflections. Quantum computing, sensing, materials science, and the fascinating interplay of quantum mechanics with biology, as explored in the preceding chapters, are testament to the theory's formidable reach.

However, there's a humility that comes with venturing into the quantum world. For every question answered, new ones arise, pushing the boundaries of what we understand and challenging our intellectual frameworks. Let's delve into some of these tantalizing open questions:

Nature of Quantum Collapse: One of the central pillars of quantum mechanics is the wave function, a mathematical entity that describes the probability distribution of a quantum system. But the act of measurement seems to "collapse" this wave function, forcing the system into a definite state. The precise nature of this collapse, and whether it's a fundamental process or merely an epiphenomenon of a deeper reality, remains a topic of fervent debate and investigation.

Quantum Gravity: While quantum mechanics and Einstein's theory of relativity are two of the most successful theories in physics, their marriage remains elusive. How does gravity, a macroscopic force, reconcile with the probabilistic and often counterintuitive quantum world? Efforts such as string theory and loop quantum gravity aim to address this, but a unified theory remains out of reach.

Quantum Consciousness: Does the human consciousness play a role in the quantum world? Some interpretations of quantum mechanics suggest that consciousness is integral to the act of measurement, and thus, the outcome of quantum experiments. While this idea is more fringe and controversial, it has led to profound discussions intersecting physics, philosophy, and neuroscience.

Quantum Entanglement and Non-Locality:

Einstein famously dubbed it "spooky action at a distance". Quantum entanglement, where particles remain interconnected irrespective of the distance separating them, defies our intuitive understanding of the universe. How can information travel instantaneously between entangled particles without violating the speed of light limitation?

Appendix

Mathematical Tools for Quantum Physics

Linear Algebra: Linear algebra forms the backbone of many quantum mechanical concepts. Matrices, vectors, eigenvalues, and eigenvectors— all vital components of quantum mechanics—are grounded in linear algebra. Transformations, rotations, and quantum operations can be represented and understood using linear algebraic methods.

Hilbert Spaces: An abstract vector space allowing for the definition of angles and distances, Hilbert spaces provide the foundation for representing quantum states.

Operators: Representing measurable physical quantities, these mathematical functions transform vectors in Hilbert spaces, essential for quantum dynamics.

Eigenvalues and Eigenvectors: An operator's eigenvectors remain in the same direction post-transformation, scaled by their corresponding eigenvalues. Measurement outcomes in quantum mechanics are directly linked to these eigenvalues.

Tensor Products: When representing states of multiple interacting particles, tensor products combine individual states into a joint state.

Probability and Normalization: Quantum mechanics interprets the squared magnitudes of complex coefficients as probabilities. For these probabilities to make sense, the state vectors must be normalized.

Fourier Transforms: This tool transitions between time and frequency domains, translating wavefunctions between position and momentum spaces.

Bra-Ket Notation (Dirac Notation): A compact way to represent vectors in Hilbert spaces, it simplifies the mathematical notation and operations in quantum mechanics.

Pauli Matrices: These are a set of three 2x2 matrices that form a basis for the space of 2x2 matrices. Used extensively in quantum mechanics, especially in the context of spin and qubits.

Commutators: Used to determine if two operators have simultaneous eigenstates. Their action is central to the Heisenberg Uncertainty Principle.

Unitary Matrices: Representing reversible quantum operations, they are crucial for quantum dynamics and quantum computation.

Wave Equation and Solutions: The Schrödinger equation is a wave equation whose solutions represent the possible states of a quantum system.

Glossary of Quantum Terms

Quantum: Relating to the fundamental units or packets of energy and matter in quantum physics.

Entanglement: A quantum phenomenon where the properties of particles become correlated and interdependent.

Wavefunction: A mathematical function describing the quantum state of a system.

Momentum: The product of an object's mass and velocity.

Superposition: The state in which a quantum system exists in a combination of multiple states simultaneously.

Photon: A quantum of electromagnetic radiation, such as light.

Hilbert Space: A mathematical space used to represent all possible quantum states of a system.

Spectral Lines: Discrete lines in a spectrum corresponding to specific energies or wavelengths.

Quantum State: The complete description of a quantum system's properties and probabilities.

Blackbody Radiation: Electromagnetic radiation emitted by a perfect absorber and emitter, like a blackbody.

Planck's Constant: A fundamental constant relating the energy of a photon to its frequency.

Further Reading and Resources

Foundational Texts:

- **"The Principles of Quantum Mechanics" by Paul A.M. Dirac:** This classic work by one of the pioneers of quantum mechanics offers a deep dive into the mathematical framework of the theory.

- **"Quantum Mechanics and Path Integrals" by Richard Feynman and Albert Hibbs:** An exploration of quantum mechanics from the perspective of Feynman's path integral formulation.

Online Resources:

- **Quanta Magazine:** An editorially independent online publication that offers insights into the latest in the realms of physics, mathematics, and computer science. The quantum section is especially enlightening.
- **The Quantum Daily:** A digital news outlet dedicated to all things quantum. From breakthroughs in quantum computing to new discoveries in quantum mechanics, this source keeps readers updated with the rapidly advancing field.

Podcasts and Multimedia:

- **"Quantum Conversations" Podcast:**
  Renowned experts discuss the latest in quantum technologies, from basic principles to commercial applications.

- **"The Joy of Quantum" YouTube Series by PBS:** A visually captivating exploration of quantum phenomena that are shaping our understanding of the universe.